Animals and their Colors

The leopard's spotted coat breaks up its shape and provides effective camouflage among rocks and trees.

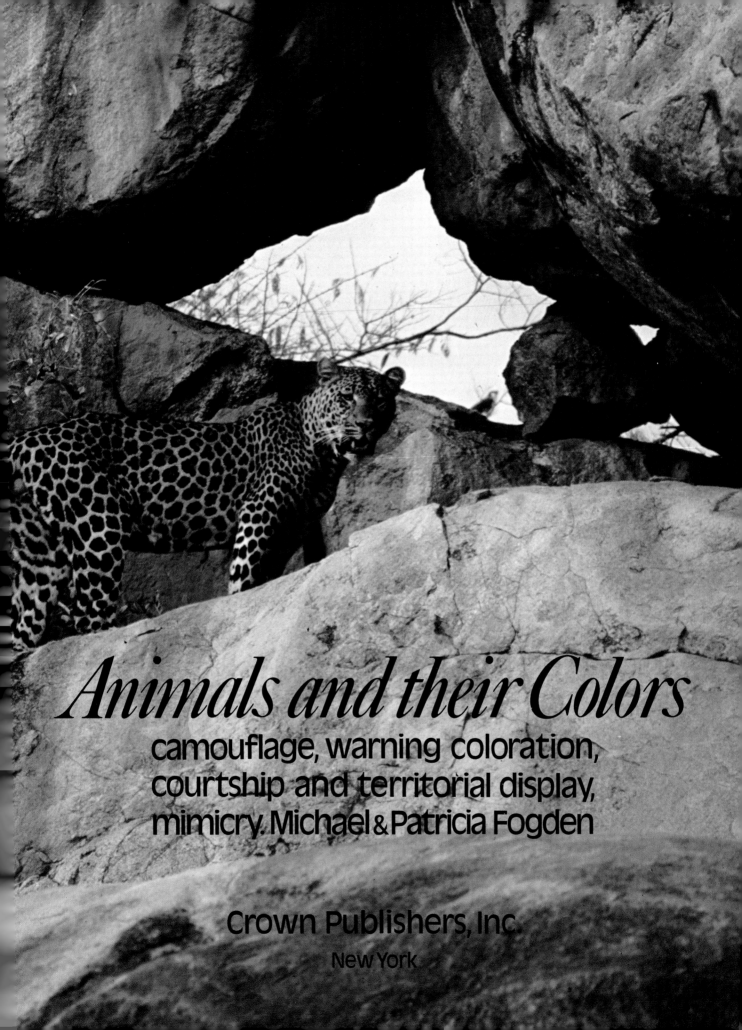

Animals and their Colors

camouflage, warning coloration,
courtship and territorial display,
mimicry. Michael & Patricia Fogden

Crown Publishers, Inc.

New York

The white winter plumage of the ptarmigan matches its snowy surroundings.

The authors are grateful to the authors and publishers
of the many books and papers to which they have
referred and from which they have quoted. These
are listed in the bibliography on page 166.

Index compiled by Susan Fogden

ISBN 0 517 514893

Printed in the Netherlands by Smeets Offset, Weert
Filmset by Trade Spools, Frome, Somerset, England

Introduction

Although animal colours is its central theme, this book is just as much about animal ecology and behaviour as about animal colours, for the three subjects are inextricably bound together. The survival value of animal colours lies in the way they adapt animals for their life in a particular environment and for their relationships with other animals.

Scientific study of the survival value of animal colours only really began in the nineteenth century, when such great naturalist-explorers as Darwin, Wallace, Bates and Müller, inspired by their experiences in the vast, unspoilt, tropical rainforests of South America and South-east Asia, developed the theories that have provided the framework for all subsequent research on the subject. Unfortunately, their theories were brought into some disrepute by overenthusiastic and uncritical writers, who tried to extend the theories to ridiculous lengths: it was even, for example, suggested that flamingos are effectively camouflaged when seen flying against a pink sunset and that the iridescent blue neck of a peacock blends with patches of blue sky visible through the branches of a tree. It was left to others, notably Sir Edward Poulton in *The Colours of Animals*, published in 1890, and Hugh Cott in *Adaptive Colouration in Animals*, published 50 years later, to review hundreds of original articles and observations and to redress the balance by the sheer weight of meticulously recorded evidence.

Later research on the subject of animal colours has delved more and more deeply, bringing to light increasingly fascinating facts and developments. At the risk of appearing unjust to others, we would perhaps single out for special mention the work of Lincoln Brower and Jane van Zandt Brower on warning colouration and mimicry, David Blest on the function of false eye-spots in insects, Wolfgang Wickler on various aspects of mimicry and Konrad Lorenz, Niko Tinbergen and their associates on colour releasers used for social signalling between members of the same species.

We are indebted to all the people mentioned above, and many others, for we have drawn freely from their work. The greater part of our book deals with camouflage, disguise, warning colouration and mimicry, colour adaptations that are involved in predator-prey relationships between different species. Unlike many writers on the subject of animal colours, we have also included a chapter which outlines some of the ways in which colours are used in social relationships between members of the same species, particularly territorial rivals, males and females and parents and young. We feel that it is important to include this chapter, for it is only possible to appreciate the full significance of animal colours when we realize that they are usually a compromise between conflicting needs for predator-prey relationships on the one hand and social relationships on the other. Finally, there is a short chapter on some of the ways man uses colour. This is a complicated subject, one that involves sociological and cultural aspects outside our own field of expertise, and we merely wish to emphasize that man makes use of colour adaptations in much the same way as other animals.

We believe the most valuable part of the book to be the collection of colour photographs, the vast majority of which are of wild animals in their natural surroundings. Although most natural history books include a few photographs which relate directly to animal colours, a reasonably comprehensive collection has never, so far as we are aware, been brought together before. We hope that they will enable readers to share our own delight in the intricacy and perfection of animal adaptations.

Michael and Patricia Fogden

Sitting among dead leaves, the woodcock provides a superb example of protective camouflage; it sits very still and its streaked and mottled, brown and buff plumage blends imperceptibly with its surroundings. The woodcock relies on its cryptic plumage even when predators, such as foxes and stoats, are very close and abandons its nest only at the last possible moment, when discovery is inevitable.

The colours of animals contribute enormously to their survival by playing an essential role in their day to day relationships with other animals. The cryptic colours of a woodcock, for example, conceal it from predators, while those of a tree-snake enable it to ambush its prey. The conspicuous colours of many frogs and insects warn of their obnoxious taste, and the brightly coloured false 'eyes' of hawkmoths, silkmoths and their caterpillars are frightening to predatory birds. Many fish assume a fine livery for courtship and territorial displays, birds recognize members of their own species by means of coloured recognition marks (often those used by bird-watchers), and nestling birds signal their hunger by gaping and displaying the brightly coloured insides of their mouths. Most of these colour adaptations function as signals for communicating information to other animals, but cryptic colours, on the other hand, obscure visual signals that might disclose the presence of an animal to its predators or prey.

The colour adaptations involved in animal relationships can be conveniently divided into two main groups: those involved in social signalling—that is communication between animals of the same species; and those involved in signalling and other interactions between different species, particularly predators and prey. Animal colours sometimes play an additional role in adapting them to their physical environment—black pigments in man and other animals, for example, absorb harmful, ultra-violet radiation—but we shall not be concerned with this aspect of animal colours.

The social use of colour signals is largely confined to animals that are active by day and have good colour vision. These include apes and monkeys, and most birds, lizards, amphibians, bony fishes and insects. However, by no means all of these see the same range of colours as man: birds poorly distinguish colours at the blue end of the spectrum, and insects are unable to distinguish reds. On the other hand, bees and butterflies can see ultra-violet colours on flowers, an ability that man is unable to match.

Apart from apes and monkeys, mammals are mainly nocturnal and have poor colour vision. Nevertheless, some species communicate to a limited extent by means of white signals that are visible even at night. Deer, antelopes, hares and rabbits, for example, have white patches on their rumps and tails which act as recognition signals. However, most nocturnal animals, together with those that live underground, in the depths of the

The brilliant warning colours and sluggish behaviour of the South American arrow-poison frogs, *Dendrobates*, advertise their poisonous skin secretions. Predators that make the mistake of molesting them associate the resulting unpleasant experience with the frogs' conspicuous and memorable colours and subsequently leave them well alone.

Right: The camouflage of the Malaysian Wagler's pit-viper is aggressive as well as protective, enabling it to ambush its prey as well as escape the notice of predators. Pit-vipers detect their warm-blooded prey by means of a pair of heat-sensitive organs situated between their eyes and nostrils. These organs are so sensitive that they can detect temperature differences as minute as 0·005°C.

sea, or in other twilight environments, depend on other types of social signals involving their senses of smell, hearing, taste and touch. Male silkmoths, for example, are attracted by the scent of females wafting from distances of more than a mile if the wind is in the right direction, and wolves locate each other by howling when they gather before a hunt.

Non-social colour adaptations are of two types. Some are involved in mutually beneficial partnerships between different animal species or between animals and plants, but the vast majority are involved in interactions between predators and prey, falling into such major categories as camouflage, disguise, warning colouration, frightening colouration and mimicry. All these adaptations are defensive, protecting animals from predators, though camouflage and disguise can be aggressive at the same time. The cryptic colours of a chameleon, for example, help it to remain inconspicuous while stalking its prey, as well as protecting it from predators. Protective and aggressive colour adaptations are most useful against animals, such as birds, lizards and fish, which are active by day and depend on sight to detect prey or predators. They are useless against animals that depend rather

on their senses of hearing, smell, taste and touch. The warning colours of many moths, for example, are no protection against bats which detect their prey by echo-location. Similarly, camouflage is no protection against shrews which secure their prey mainly by smell and touch. On the other hand, camouflage, warning colouration and other colour adaptations can be equally effective when seen in monochrome by nocturnal animals, notably bush-babies, tarsiers, civets, cats and owls, which depend on their enormous eyes to a considerable extent. Many animals abroad at night are cryptically coloured in browns and greys for this very reason and there are even a few examples of nocturnal warning colouration, notably skunks and the zorilla.

The colours of animals frequently portray a conflict between their requirements for social signalling and their requirements for interactions with predators and prey. Sometimes this conflict has been resolved entirely in one direction. Many birds, for example, are brilliantly coloured almost entirely for social purposes; they can afford to be so because they breed in safe places where cryptic colours are unnecessary, and rely at other times on their alertness and flying ability to elude predators. Other birds are completely cryptic in their colouring and have to utilize other types of social signals, such as song. It is very noticeable that the best bird songsters are drably coloured, while brilliantly coloured species generally have drab songs or no song at all to speak of. Some birds have the best of both worlds by combining a basically cryptic

As a second line of defence, the normally cryptic eyed hawkmoth suddenly reveals staring false eyes which are frightening to birds and most other small predators.

The white markings of many nocturnal animals are visual signals which function even in the dim light provided by the moon or stars. They act as either warning signals or social recognition signals. In the case of the badger, a formidable adversary, they probably serve both purposes.

The colours of many animals play an important role in their social lives, particularly in territorial and courtship displays. Male black grouse, for example, congregate on traditional arenas or leks to challenge each other and show off their splendid plumage. Their impressive displays intimidate rivals and attract the soberly coloured females that visit the lek to mate.

Right: Like bush-babies, owls and a number of other nocturnal predators, the Bornean tarsier has enormous eyes and hunts mainly by sight. To avoid predators such as these, rodents, lizards, insects and other small animals need to be protectively coloured even at night. The tarsier's eyes are so sensitive that the pupils are contracted when it begins to hunt at dusk and only expand fully in the near total darkness of rainforest at night.

The melanistic form of the peppered moth is rare in areas where tree-trunks are covered in lichens, because it is conspicuous to predators on such a background. It survives well, however, in areas where pollution has killed off lichens and blackened tree-trunks; during the last century it has provided a striking example of natural selection in action by replacing the typical lichen-like form in most of the industrial areas of Britain.

plumage with concealed areas of colour which can be exposed in courtship and other displays. Similarly, many butterflies have beautiful social colours on the upperside of their wings which are visible in flight and cryptic colours on the underside which conceal them when they alight and close their wings.

However, the conflict is not so easily resolved in animals whose colour requirements differ according to whether they are immature or adult, male or female, breeding or not breeding, or because they live in variably coloured surroundings. Such conflicting requirements have resulted in the evolution of different colour forms in some animals (which are said to be polymorphic) and the ability to change colour in others. Many desert animals have colour forms which tone with the reds, browns, ochres and greys of different desert soils, while cuttlefish, octopi and chameleons can change colour at will to match different backgrounds or to produce colourful social displays.

Sometimes natural selection acts rapidly producing colour forms to match changing environmental conditions, a striking example being industrial melanism in the peppered moth and other Lepidoptera. In Britain prior to 1850 the peppered moth normally existed only as a pale form matching the lichen-covered tree-trunks which are its natural resting place, though a rare black form had been recorded occasionally. By 1900 the black form had become common and the pale form rare in areas where industrial pollution had killed off lichens and blackened tree-trunks. In Manchester, for example, the black form accounted for 95 per cent of the population. In 1953 H. B. D. Kettlewell proved conclusively that bird predation was the selective agency responsible for this remarkable change. He released large numbers of both colour forms in two areas, one rural and one industrial, where they were preyed upon by wild birds. In each case, he found that the contrasting form was predated in greater numbers, more of the black form being taken from lichen-covered trees in rural areas, and more of the pale form from soot-covered trees in industrial areas. The preponderance of the pale form in rural areas and the black form in industrial areas has persisted almost to the present day, but there are now signs of a return to the pre-1850 situation, as a result of cleaner industrial areas, smokeless zones, and the consequent return of clean trees. These changes offer a remarkable example of how protective adaptations are refined by the interactions of predators and prey.

The nature of animal colours

The pink colour of the plumage and legs of the lesser flamingo (and other flamingo species) is due to a carotenoid pigment obtained directly from the microscopic algae which it sieves from the alkaline waters of African soda lakes.

A rainbow vividly demonstrates how white light can be split into red, orange, yellow, green, blue, indigo and violet spectral colours, each of which is composed of different wavelengths of light. When sunlight falls on a coloured object some of the spectral colours are absorbed or cancelled out, so what we see is the complementary colour, which is a mixture of the remaining unabsorbed colours. When sunlight falls on a goldfish, for example, the violet to green wavelengths are absorbed by a pigment while the complementary red to yellow wavelengths are reflected and seen as orange. The fur, feathers and skin of most animals have some colour producing property, which may be pigmentary, structural, or a combination of both. Pigmentary colours are caused by *chemical* compounds which selectively absorb and reflect particular wavelengths of light, while structural colours are caused by the *physical* nature of a surface. When all the light wavelengths are absorbed, the result is black; when all the light wavelengths are reflected, the result is white.

PIGMENTARY COLOURS

Animal pigments are generally granules or droplets embedded in skin, fur or feathers, or in fatty tissues or specialized cells visible below the skin. A few pigments have a physiological function, notably haemoglobin which is involved in oxygen transport in the blood of vertebrates, but the majority can probably be thought of as waste products from a physiological point of view. Of course, their biological importance can hardly be overestimated.

Most of the blacks, browns, reds and yellows, and some greens and blues, in the animal kingdom are produced by pigments. Pigments called melanins are almost universal, being responsible for blacks, most browns and even some of the duller reds and yellows, such as the red hair of humans and the yellow downy plumage of chicks. Bright reds, oranges and yellows are generally caused by carotenoids (so called because one of them—carotene—is found in carrots) or pterins. Carotenoids are important pigments in most animal groups, while pterins occur mainly in insects such as Pierid butterflies and wasps.

Animals synthesize most pigments, but not carotenoids which herbivores obtain directly from plants and carnivores at second hand from herbivores. Insufficient carotenoids are often the reason for the dull colours of birds in captivity. Thus

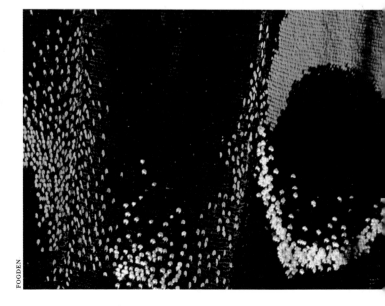

The brilliant blue of the feathers on the back of the African pygmy kingfisher is a structural blue caused by Tyndall scattering. Minute air bubbles in the feathers scatter and reflect the blue wavelengths of light, which are then visible against a background of black melanin granules. The structural nature of the colour is proved if the feathers are viewed by transmitted light: their blue colour disappears and they appear dull brown. The chestnut of the underparts is caused by a pigment, probably a melanin.

Above, left: The basic colour of the orange-tip butterfly is a structural white; its markings are due to scales containing brown melanin or orange or yellow pterin pigments. Brown and yellow scales are interspersed on the hindwings to give the appearance of an olive-green colour. Pterins, like melanins, are manufactured, not obtained ready-made from plants.

Left: The individual scales on the hindwing of the African swallowtail butterfly, *Papilio lormieri*, are clearly visible when magnified ×8. The dark brown, orange and yellow scales contain melanin and pterin pigments, while the blue scales owe their colour to Tyndall scattering.

Melanin pigments are responsible for most mammalian colours. The black, brown and russet colours of the bushbuck, for example, are due to melanin granules deposited in its hairs. White is due to light being scattered and totally reflected by air bubbles in hairs devoid of pigments. By day the bushbuck lies up in thickets where its patterned coat blends well with its sun-dappled surroundings; colour resemblance is less important than tone and pattern, because the bushbuck's main predators are lions and leopards, which see only in monochrome.

The blue sheen on the transparent wings of many damselflies is caused by Tyndall scattering; it is visible only from certain angles and against a dark background.

caterpillars and grasshoppers are a mixture of blue and yellow pigments, while the olive-greens of many birds, such as tits and white-eyes, are a mixture of black and yellow, the black usually being in the barbs of the feathers, the yellow in the barbules. The green on the underwing of the orange-tip butterfly is produced by brown and yellow scales interspersed in much the same way as Impressionist painters produced green with tiny, interspersed dabs of black and yellow paint.

The dominant green plant pigment—chlorophyll—which is responsible for the greenness of grass and leaves, is not synthesized or utilized by animals, though it can sometimes be seen in the gut of semi-transparent caterpillars, and also occurs in algae living on the shaggy hair of sloths, giving them a greenish tint. The algae, incidentally, are fed upon by the caterpillars of the sloth moth, providing a curious example of the way in which extremely specialized habitats are sometimes successfully exploited.

NON-IRIDESCENT STRUCTURAL COLOURS

Non-iridescent structural colours in animals are caused by the scattering of light by small particles. If the particles are sufficiently small they scatter more short (blue and violet) wavelengths of light than long (red) wavelengths. The result is blue reflected light and red transmitted light, an effect visible when the short wavelengths of sunlight reflected from the earth are scattered by minute dust particles suspended in the atmosphere and reflected back to earth. Viewed against the black background of space this results in the brilliant blue of the sky. The blue colour of the eyes of animals, including those of humans, can be explained in the same way. Minute protein particles in the stroma of the iris scatter the short wavelengths of light and the reflected blue light is viewed against a black melanin pigment layer at the back of the iris. Albinos have pink eyes because they lack a melanin background, so that the blue colour is masked by blood in the capillaries at the back of the iris. In brown eyes the blue colour is masked by the melanin pigment granules embedded in the stroma among the scattering particles. The essential ingredients for the production of this blue structural colour, known as Tyndall blue, are a scattering layer and a black background. Tyndall scattering is responsible for the brilliant blue plumage of such birds as kingfishers and parrots, for the blue face and genitals of mandrills and

flamingos in zoos only remain pink if fed a carotenoid-rich diet during their moult. Similarly, a pet wrinkled hornbill in Sarawak lost the normal bright orange colour of its bill and only regained it after being fed plenty of tomatoes and egg-yolks. Carotenoids are often soluble in animal fats and occur, for example, in butter, the yellowness of which depends on the amount of carotenoids eaten by cows.

Many greens and blues in animals are wholly or partly structural, though some shades, particularly the duller ones, are caused by pigments. Carotenoids linked with a protein are responsible for many blues and greens, including the dark blue of live lobsters. The red colour of a cooked lobster is the result of the protein link being broken down, leaving only a red carotenoid. The greens of many vervet monkeys, and for many brilliant blues throughout the animal kingdom.

The brilliant green of many birds, snakes, lizards and frogs is produced by a Tyndall blue in combination with a yellow pigment filter, through which light passes both before and after being scattered. Here there are essentially three functional layers producing the colour—a yellow pigment layer, a scattering layer and a melanin pigment layer against which the scattered Tyndall blue is viewed. Three layers of this type are responsible for the green colour of wild budgerigars, and domestic colour varieties have been bred by selecting for genes which prevent the accumulation of the pigment layers. Lack of the yellow pigment results in a blue variety, lack of the black pigment in a yellow variety, while white budgerigars are lacking in both pigments.

The yellow pigment associated with Tyndall blue in the production of green colours is often soluble in alcohol. For this reason specimens of green snakes, lizards and frogs preserved in alcohol are sometimes blue. This has occasioned the odd misnomer: a brilliant green Australian tree-frog, for example, was described and named *Hyla coerulea* (blue tree-frog) by a museum worker who only saw it as a pickled specimen. In birds the yellow pigment is sometimes liable to fade in bright sunlight. The green magpies of Malaysia and Indonesia, for example, turn turquoise blue when they leave the gloom of primary forest to live in young, sunlit, secondary forest.

WHITE AND GREY

Whiteness in animals is also a structural colour and is caused by the total reflection or scattering of light by particles too large to scatter only the short blue wavelengths. For example, the whiteness of hair is caused by scattering from minute air spaces, and that of moths and butterflies by reflection from the complicated ribbed and grooved surfaces of their scales. Pierid butterflies have a white pterin pigment, but apparently also owe their whiteness to structural scattering, for it is retained even when the pigment is dissolved.

Greyness in animals is often caused by whiteness being partially masked by a melanin pigment, or by a mixture of completely dark and completely white elements. Grey human hair, for example, results from white hairs being interspersed with dark hairs, and the grey feathers of many birds result from mixing white barbules with black.

IRIDESCENT COLOURS

Anyone who has watched ducks in the wild or on ornamental ponds will have noticed that the sheen on the head of a drake mallard changes from metallic green to metallic purple depending on the angle from which it is viewed. Similar iridescent colours occur in all sorts of animals, ranging from golden moles to cuttlefish and earthworms, but they are particularly beautiful and conspicuous in birds like pheasants, hummingbirds and sunbirds, and in many butterflies and beetles. All iridescent colours in animals are structural; the majority are caused by interference of light, but diffraction is responsible in a few cases, notably in the rainbow colours of Ctenophores, the golden bristles of the sea-mouse, and the mother-of-pearl colours of mollusc shells.

Interference colours are perhaps best understood by considering the familiar rainbow colours of a thin film of petrol on a puddle of water. Light falling on the film is reflected from the surface of the water as well as from the surface of the petrol. Because of the distance they travel through the film, some wavelengths reflected at the water surface do not remain in phase with equivalent wavelengths reflected from the petrol. The out-of-phase wavelengths vibrate in opposite directions, interfere with each other and cancel each other out, leaving only the complementary wavelengths, which are seen as an appropriate colour. The surface of the film appears rainbow-coloured because the distance travelled through the film, and hence the colour produced, varies according to the angle from which it is viewed. The interference colours of animals are produced in an essentially similar manner, the necessary thin films being in the form of minute platelets in cells, or in the keratin or chitin of fur, feathers, scales and cuticle.

COLOUR CHANGES

Many animals have the ability to change colour by one means or another. Some colour changes are simply a part of the process of becoming adult, as for example in gulls losing their brown juvenile plumage and turning white. Others are a regular seasonal occurrence associated with breeding or

The brilliant green of the Malaysian green broadbill is caused by a structural Tyndall blue seen through a yellow pigment filter. Most brilliant greens in the animal kingdom are formed in this way, but dull greens are usually due entirely to pigments.

changing ecological conditions brought about by the alternation of summer and winter or wet and dry seasons. Thus many birds don a breeding plumage, and many arctic animals assume a white coat in winter. Colour changes in mammals and birds are mainly of these types and are achieved relatively slowly by moulting at regular times of the year. More rapid colour changes are impractical in mammals and birds because of their covering of hair or feathers, though a few species with bare facial skin have the ability to flush pink or red. Good examples are turkeys, vultures and humans. The colouring is caused by the red pigment, haemoglobin, becoming more prominent when extra blood is forced into capillaries underlying the facial skin. In man a red flush is associated with emotions such as anger or embarrassment. Blanching—the opposite of flushing—is caused by blood being drained from the face and is associated with anger, fear or shock.

Though the ability to change colour rapidly is rare among mammals and birds, it is widespread in the rest of the animal kingdom. It enables animals to match the colour of their background closely, and to be flexible in their choice of surroundings, as well as to signal changing emotions. Some colour changes are remarkably rapid and equally rapidly reversible; others occur more slowly and have a different physiological basis.

The speed with which chameleons can change colour is proverbial, but the most rapid and varied colour changes are those of Cephalopod molluscs, such as cuttlefish and octopi. The changes are accomplished by means of specialized cells called chromatophores which contain various pigments. Each chromatophore is bounded by an elastic membrane and its size and shape is controlled by radial muscle fibres. When the muscles contract the membrane is pulled into the shape of a flat disc and the pigment is spread into maximum prominence; when the muscles relax the membrane contracts and the pigment is compressed into a tiny inconspicuous sphere. By expanding and contracting appropriate chromatophores an octopus or cuttlefish can send shimmering colours flickering across its body in a display which far surpasses the colour changes of all other animals.

The iridescent blue, green and bronze colours of the peacock are caused by interference of light, in the same way as the colours on a thin film of petrol or a soap bubble. Iridescent colours vary with the viewing angle, so, by quivering its gorgeous tail, the peacock can display a shimmering array of colours.

BLOSSOM/NHPA

In spring, as the northern snows melt, the varying hare moults from a white, winter coat into a brown, summer one. Many other mammals and birds undergo regular seasonal colour changes by moulting their fur or plumage. Some, like the varying hare, alternate between two cryptic dresses which fit different environmental conditions, others between a cryptic dress and a colourful breeding dress.

The East African *Chamaeleo dilepis* and other chameleons have several types of chromatophores (colour cells) in their skin, but effect major colour changes solely by expanding or contracting black melanin pigment within large branched cells. Chameleons change colour to some extent to match different surroundings, but show their most striking colour changes in territorial, courtship and frightening displays. This one, for example, becomes most vividly coloured when displaying to other chameleons and almost black when attacked by predators. Similar rapid colour changes are found in many reptiles, amphibians, fishes and invertebrates.

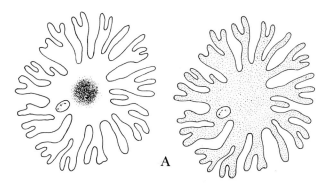

A

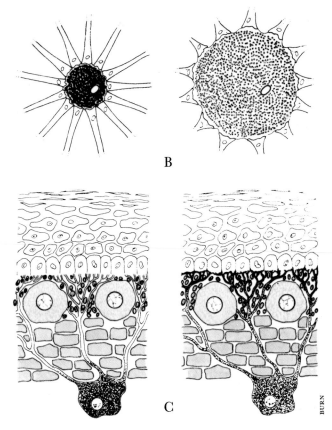

B

C

BURN

(a) Rapid colour changes in reptiles, amphibians, fish and most invertebrates are due to the movement of pigments within special branched chromatophores. The pigments can be concentrated at a single point or spread throughout the branches, their dispersal being governed by visual stimuli.

(b) Rapid colour changes in octopi, cuttlefish and other Cephalopod molluscs are brought about by complex chromatophores which differ from those of other animals. Each chromatophore consists of an elastic, pigment-filled cell surrounded by muscle cells. The central cell is small, spherical and unobtrusive when the muscles are relaxed, but is pulled into a flat, coloured disc when the muscles are contracted. Cephalopods have variously coloured chromatophores which can expand or contract independently, giving an enormous range of possible colour variations.

(c) The skin of many lizards and frogs has three main colour producing layers: a superficial layer of oil droplets containing yellow carotenoid pigments; a layer of cells containing minute scattering particles which produce Tyndall blue; and a deep-seated layer of melanophores with branches extending into the other two layers. The colour of the lizard or frog varies from pale yellow, through various shades of green and brown, to dark brown. Its shade depends on whether the black pigment in the melanophores is fully contracted, expanded sufficiently to provide a black background to make the Tyndall blue visible, or expanded into the branches so as to obscure or partially obscure the Tyndall blue and yellow pigment layers.

Nevertheless, the abilities of many reptiles, amphibians, fish and arthropods are considerable. Their chromatophores operate differently from those of Cephalopods, being in the form of branched cells within which the distribution of pigment can be changed. It can be spread throughout the cell to produce a strong colour or concentrated to a single point where it is inconspicuous. Though many fish have chromatophores containing several different pigments, the colour changes of chameleons, lizards and frogs are brought about mainly by black melanophores. The way in which this happens is well illustrated by many frogs and lizards which have a basic green colour produced by a Tyndall blue in combination with a yellow pigment filter. The scattering layer which produces the Tyndall blue is backed by melanophores whose branches extend outwards into the scattering and yellow pigment layers. The shade of green darkens when black pigment expands into the branches and mixes with the blue and yellow, and turns to brown when the blue is completely obscured. Some species can become almost black and tend to do so when angry or frightened. The melanin pigment can also be completely retracted, making the animal appear pale yellow, as the Tyndall blue is invisible without a black background. This is what happens in chameleons and many tree-frogs at night, and it makes them very conspicuous in the light of a torch.

Not all colour changes among invertebrates and the lower vertebrates are so quick. Some take several weeks and are brought about by gradual changes in the types and quantities of pigments present in the skin. New pigments are synthesized or accumulated from food, and old ones broken down or excreted. Particularly good examples of such changes are afforded by crab spiders and praying mantids, which habitually lurk in flower heads where they prey on visiting insects. Crab spiders in particular regularly alternate between yellow and white flowers and can gradually change their colour accordingly.

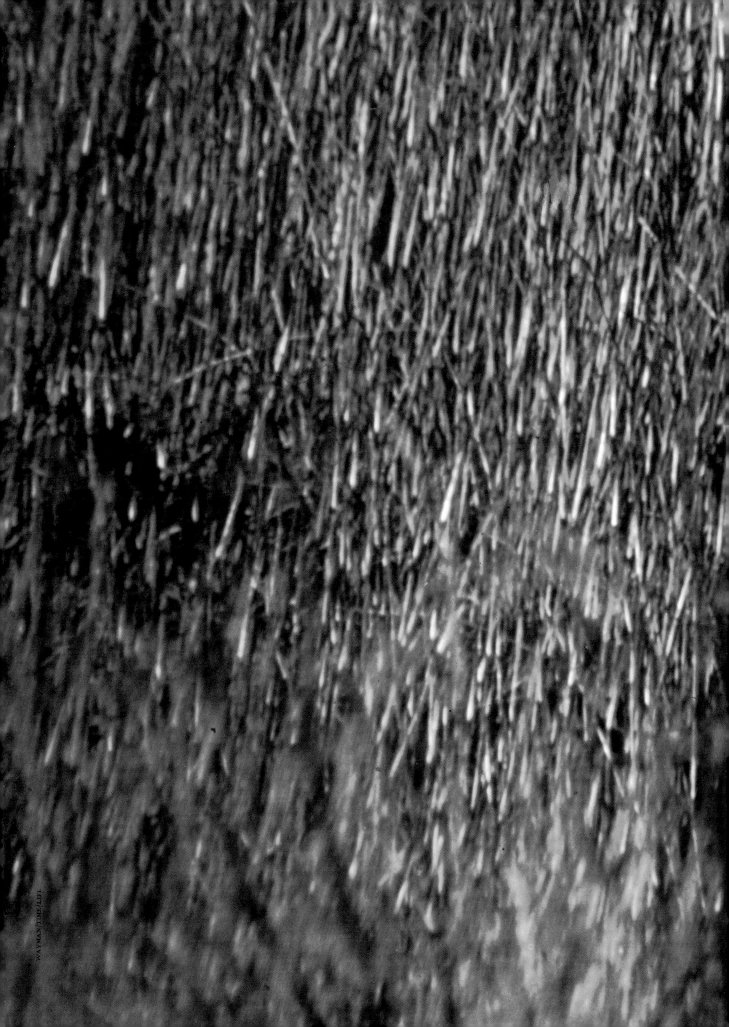

Camouflage and disguise

The tiger's stripes provide an effective camouflage in dry, brown grass.

Concealment by camouflage or disguise is essential to the life of many animals. It can be either protective, as in animals seeking to escape the notice of predators, or aggressive, as in predators seeking to ambush their prey. Frequently concealment serves both functions in the same animal, since many small predators are preyed upon by other predators more powerful than themselves.

In discussing cryptic colouration it is usual to distinguish between camouflage and disguise, though it must be said that the two categories merge so imperceptibly that it is sometimes difficult to assign a cryptic animal to one or the other. Nevertheless, the distinction is useful: camouflage breaks up the shape and outline of an animal, making it merge with its background, whereas disguise makes an animal look like some specific but inanimate part of its environment, such as a leaf or twig, which is of no interest to predators and which registers no alarm in prey. In other words, camouflaged animals rely on effacing themselves, whereas disguised animals rely on not being recognized for what they are, even though their outline is in full view and quite easily seen.

To be fully effective camouflage and disguise must be combined with appropriate behaviour. A cryptic animal must be still, it must choose suitably coloured surroundings, and it must sit in a suitable posture. Immobility is of paramount importance, since even the most marvellous camouflage or disguise will be spoilt by movement. For this reason protective camouflage and disguise are most highly developed in animals which feed by night and rest in exposed positions by day, and are of little use to active diurnal animals which rely on other methods for their protection. Most moths, for example, are active at night and are almost invisible on the tree-trunks and dead leaves where they rest by day, whereas most birds are diurnal and rely on alertness and flight to escape predators. Similarly, aggressive camouflage and disguise are most highly developed in predators, such as cats, chameleons, many fish and praying mantids, all of which lie in wait for their prey, and are not developed at all in animals such as the hunting dog which relies on speed and endurance to run down its prey. When cryptic animals do have to move they do so with great stealth. For example, chameleons and praying mantids move with a rocking gait which resembles the movement of leaves in a gentle breeze, the leaf-fish drifts towards its prey, and stalking lions use all available cover and freeze immobile whenever their prey looks in their direction.

HOSKING

BLACKBURN/NHPA

Left: The desert lark has a dark grey-brown race which lives on a ridge of grey lava in the Jordan Desert. A pale, buff race inhabits the surrounding desert and other colour forms occur elsewhere in the Middle East and North Africa. Each colour form matches the colour of the soil and rocks where it lives.

Below, left: The European nightjar and its young illustrate the way in which camouflaged animals merge so perfectly with their background that their shape is hardly discernible. To make the nightjars' camouflage even more perfect, their potentially conspicuous eyes are closed to narrow slits.

The leaf-mimicking shorthorn grasshopper, *Trigonopteryx*, from Borneo, illustrates the way in which disguised animals differ from those that are camouflaged. Its outline does not merge with its surroundings; it is perfectly obvious, but appears to be the outline of a totally different object—a half-eaten leaf. Its disguise is enhanced by its leaf-like posture among half-eaten leaves.

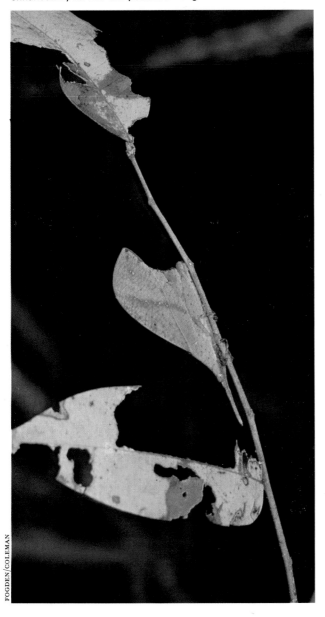

FOGDEN/COLEMAN

Not only should cryptic animals remain still and seek appropriate surroundings, but they should also be well spaced out. There is evidence that predators form a searching image in their minds, which gets stronger the more frequently they encounter a particular species. In other words, a predator gets its 'eye in' for a particular prey, thereby increasing the efficiency with which it catches it. Experiments have shown that individual birds which have developed a very strong searching image for certain prey will, at least for a while, overlook alternative prey which has become more abundant. Spacing out is clearly a useful strategy for cryptic animals, not only because it ensures that the searching images of predators remain relatively weak, but also because it increases the chance that predators will encounter and form searching images for other less well spaced prey.

RACIAL COLOUR FORMS

In a wide range of cryptic animals the need for harmony with their surroundings has led to the evolution of local colour forms or races. Apart from the peppered moth, and other cases of industrial melanism, the most famous example is that of the desert lark, which ranges in colour from dark grey to near white, depending on the colour of the rock and sand on which it lives. Near the Azraq Oasis in Jordan a dark grey-brown race lives side by side with a pale race, the one on a low narrow ridge of grey lava, the other on the surrounding pale ochre desert. The colour variations found in the desert lark are paralleled by many desert mammals, snakes, lizards and insects. In the desert of New Mexico, for example, a very dark race of the pocket mouse, (*Perognathus intermedius*), lives on a black lava flow in the Alamogordo region, while in the more typically coloured surrounding desert there are buff and tawny forms. Not far away in the White Sands National Monument—a desert of pure white gypsum—there also lives an almost pure white pocket mouse, (*P. gypsi*), found nowhere else in America.

OTHER COLOUR FORMS

Insects quite often have two or more cryptic colour forms within the same population of a species, or even living on the same plant. In gardens in Sarawak it is quite easy to find green and brown forms of a small stick-insect on the same bushes, the green form resting among green foliage, the

33

brown on clusters of dead leaves caught in the branches. By marking individuals we confirmed that they remained basically either green or brown, though both forms altered their shade slightly. This type of polymorphism is widespread in insects, particularly stick-insects, grasshoppers and praying mantids. It is probably a strategy that enables them to increase their numbers without increasing the likelihood of predators encountering them often enough to form a strong searching image for either.

Alternatively, colour forms can be adapted to live on different food plants. The stick-caterpillar of the peppered moth, for example, feeds on birch or oak, the twigs of which differ in colour, texture and markings. Sometimes the twigs are covered in lichens. The caterpillar therefore has three appropriate forms, being a smooth purplish-brown when on birch, brownish-green when on oak, and a mottled green when on lichen-covered twigs. In each case it exactly matches the texture and markings as well as the colour of the twigs upon which it rests. Probably each of these forms is genetically

These green and brown stick-insects are different colour forms of the same species and live together on bushes and trees in Sarawak. Predators need different searching images for the two forms and are therefore likely to overlook one form when they have their eye in for the other.

34

determined, which form actually develops depending on what the caterpillar perceives through its eyes when it hatches from its egg. What it sees operates a 'switch mechanism', bringing into action appropriate genes, after which the colour form is irreversible. Similar colour adaptations are found in pupae of cabbage white butterflies which match backgrounds ranging from pale green to mottled grey. In this case it has been proved that the colour of pupae is determined by the background colour seen by the fully grown caterpillar just prior to pupating. If caterpillars are blinded their pupae are always green.

COLOUR FORMS OR CHANGES CAUSED BY DIET

In some cases the different colour forms of caterpillars result from their diet. Caterpillars of pug moths, for example, match the colour of the flowers on which they feed—blue scabious, pink thistles, or yellow composites—by accumulating their food

The caterpillars of a common Malaysian hawkmoth, *Rhyncholaba*, feed on the variably coloured leaves of an ornamental aroid, *Caladium*. All remain green until their final instar, when those feeding on pink leaves turn pink to match their surroundings—a great advantage when they are so large.

35

plant's own pigments. In Sarawak caterpillars of a hawkmoth, *Rhyncholaba*, are very common in gardens, feeding on the enormous green or pink and green, ornamental leaves of an aroid, *Caladium*. All the caterpillars remain green until their final instar, but at that stage individuals feeding on pink and green leaves become a matching pink. The colour change occurs only then and not before because the red pigment in the aroid is a fat-soluble carotenoid, and it is only during their final instar that caterpillars accumulate a fat layer in which the pigment can dissolve. The colour change is an advantage at this stage because four-inch-long green caterpillars would be conspicuous on pink leaves.

Other examples of variable colour resemblance linked with diet are found in the shell-less Nudibranch molluscs known as sea-slugs. Species in several genera (*Doris*, *Archidoris*, *Rostanga* and others) accumulate pigments from the sponges on which they rest and browse, and exactly match the sponges' varied and beautiful colours—green, red, salmon-pink, yellow or creamy white. However, it is debatable whether this is a truly cryptic resemblance, for sponges are well protected against most predators by the needle-like spicules embedded in their skin. The sea-slugs are among the few animals that can feed on them, and they actually incorporate the spicules intact in their own skin. Thus they too are well protected against predators and have little need to be inconspicuous.

SLOW COLOUR CHANGES

Many invertebrates are able to adjust their colour to new surroundings by gradually changing the types and quantities of pigments in their skin. Such changes are usually reversible and can take anything from a day or two to several weeks to accomplish. The colour changes of crab spiders, for example, take from five to twenty days, and are made necessary by the short flowering seasons of many of the flowers in which they wait for prey.

Having hatched from the same egg mass, these green and red nymphs of the African flower mantis, *Pseudocreobotra wahlbergi*, dispersed onto flowers of different colours and manufactured appropriate pigments to match their chosen background. Their capacity for colour resemblance is limited, however, to red, green and intermediate shades. Flower mantids use their chosen flowers as lures, preying on the insects that are attracted by the flowers' nectar and pollen.

The North American species *Misumena vatia* changes colour from white to yellow when it moves from ox-eye daisies or white fleabane to golden rod in July or August. Other variable flower-haunting species are two beautiful flower mantids—*Pseudocreobotra wahlbergi* of Africa and *Hymenopus coronatus* of Malaysia and Indonesia. On one occasion in Uganda nymphs from a single egg mass of *P. wahlbergi* dispersed onto the pale greeny-white flowers of a weed and bright pink blossoms of oleander, and gradually accumulated appropriate pigments to match their surroundings. Nymphs that were swapped around between flowers gradually changed colour to match their new surroundings, but remained an intermediate pink-green shade for several weeks. Similar changes occur in *H. coronatus* which is mauve or pink on flowers of *Melastoma* or orchids, but a pearly-white on the blossoms of ornamental gardenias and frangipani in gardens. Among other insects there are striking colour changes in shorthorn grasshoppers, though the changes are slow and often involve moulting the skin. *Acrida turrita* of Europe and the Middle East varies over a particularly wide range of colours—green, yellow, rufous, brown, grey and almost black—depending on the colour of its surroundings. Similarly, many African insects develop dark forms in areas blackened by grass fires.

Among vertebrates examples of slow colour changes are particularly numerous in fish. Flatfishes, such as plaice, brill and turbot, which are capable of very rapid colour changes involving movements of pigments within chromatophores, also undergo slow reversible changes involving the actual number and distribution of chromatophores. Melanophores are least numerous when the fish live on pale sand in shallow water, and increase in number when they move to dark muddy areas or deep water. Though these changes are reversible in the laboratory, they are normally associated with a change in behaviour at a particular life stage. Young plaice, for example, live in well-illuminated inshore waters with a sandy bottom and migrate to more varied areas in deeper water as they grow older. Minnows are also adept at matching darker or lighter surroundings, a fact well known to fishermen who keep live minnow bait in white containers to make it a more conspicuous lure for river fish. Much more striking colour changes occur in a blenny from the coast of California, which can change from dull red to olive-green or olive-yellow, depending on the colour of the seaweed among which it is living.

MORRIS/ARDEA

Cuttlefish and other Cephalopod molluscs are able to adapt their colour to different surroundings more rapidly than any other animals. The cuttlefish has a normal swimming pattern of flickering, dark, transverse stripes which are said to be cryptic when viewed from a distance.

The American horned lizard can vary its colour to blend with the tones and textures of the desert soils where it lives. Having just been moved from an area of pale, granitic soil, one of these two individuals is very conspicuous, but it will match its new reddish-brown background within a few hours.

ROSS

RAPID COLOUR CHANGES

The colour changes of octopi, cuttlefish and squids require only two-thirds of a second, being the most rapid colour changes of all animals and much quicker than those of chameleons. Their most spectacular displays are concerned with territorial defence and courtship, though their colour changes concerned with camouflage are also remarkable. An octopus, for example, can quickly match background colours ranging from white to black, or reproduce the mottled effect of a sandy sea bottom. A cuttlefish can match a strongly contrasting pattern of light and shade with a disruptive contrasting pattern of white bands on a dark background. It has even been said that cuttlefish can simulate the play of light and shade from moving

waves above. Certainly they can send waves of colour rippling over their backs, which may make them less conspicuous in well-lit surface waters.

Plaice, turbot and other flat-fishes can make rapid colour changes in addition to the slow ones already mentioned. They are particularly adept at making rapid adjustments to match the texture of their background, and can harmonize with equal ease with mud, sand or a coarse gravel seabed. Many other fish make rapid colour changes, some of which concern camouflage. The Nassau grouper, for example, can match any of the varied backgrounds in the sea-caves, crevices and reefs in which it lives.

Everyone knows that chameleons can change colour within minutes, but it is not so well known that many lizards and frogs share this ability. In

fact, the colour changes of many arboreal lizards, such as *Calotes* in Asia and *Anolis* in America, are so rapid that they are often mistakenly called chameleons. In fact, they have much the same range of colours as chameleons—mainly various shades and mixtures of green and brown. Similarly, the terrestrial horned lizard of the deserts of the American south-west is able to match any of the varied brown, yellow and grey colours which dominate its surroundings.

The colour changes of frogs and toads are no less striking. An often repeated story in books on camouflage tells of a lady who made pets of three toads which lived in her garden—a green one among her rhododendrons, a speckled brown one in her rockery, and a dark olive-brown one on the mud by her stream. Only when she recognized that they had certain mannerisms in common did she discover that her three toads were really only one. The colour changes of most frogs and toads are restricted to shades ranging from pale green or yellowish-green to dark green or brown, but *Phrynobatrachus plicatus* from West Africa can be white, gold, pink, brick-red, maroon or purple, in addition to the more usual greens and browns. The colour changes are sometimes very rapid. In Sarawak, for example, we often came across a common tree-frog, *Rhacophorus*, which changed from bright green to brown within seconds when frightened from a green leaf to the brown leaf litter on the forest floor. In both positions it was most inconspicuous.

SEASONAL COLOUR CHANGES

The most striking seasonal colour changes occur in Arctic and sub-Arctic mammals and birds, such as foxes, stoats, weasels, hares, lemmings and ptarmigan. All have brownish or greyish coats in summer which match the lichen-covered tundra in which they live, but moult into a white coat in winter to match the winter snows. There is good evidence that these changes are adaptive for the colour of several species remains unchanged in areas where there is little snow. The mountain hare, for example, almost always moults into a white coat in Scandinavia, generally does so in Scotland, but hardly ever does in the milder climates of England and Ireland. Much the same is true of stoats and weasels. In the far north there are a number of species which remain white all the year round, notably the polar bear, American polar hare and the snowy owl, and this is an obvious advantage in areas where patches of snow and ice remain throughout the year. On the other hand, it is perhaps less obvious why some species, such as the reindeer, musk ox, wolverine and raven, should retain a dark coat throughout the winter. The answer lies in their ecology and behaviour: reindeer and musk oxen are large gregarious animals

Right: By being white in winter and dark brown in summer, the North American snowshoe or varying hare matches the prevailing colours of the sub-Arctic forests where it lives. Its cryptic colours help to conceal it from predators.

The Malaysian tree frog, *Rhacophorus*, can change colour within seconds to match green foliage or dead brown leaves on the forest floor. At night, when camouflage is less essential, it becomes a pale translucent yellow (perhaps to attract the nocturnal insects that visit pale-coloured flowers at night).

not dependent on being cryptic to escape predators, while wolverines and ravens are mainly scavengers which do not have to hunt for prey.

Another type of seasonal colour variation is found among some of the deer living in deciduous forests. In summer, both the fallow deer and the sika have brown coats spotted with white, the white spots resembling dappled sunlight shining through leaves. In winter, when the trees are bare of leaves and the light is uniform, their coats are plain grey or brown. It is significant that a race of the sika that inhabits the evergreen forests of Taiwan retains its dappling of white spots throughout the year.

In areas where there are distinct wet and dry seasons many insects have seasonal forms which differ in colour in yet another way. The colour changes occur in successive generations, not in successive coats of the same individuals. Most seasonal changes of this kind result in colour forms which match the prevailing vegetation, which is green in the wet season and brown or yellow in the dry season, but there are many exceptions to this rule among African butterflies. Several species of pansies, *Precis*, for example, have a wet season form that is conspicuous and a dry season form that is cryptic, especially when at rest. The reason for these changes has never been fully explained, but it is probable that the species concerned are slightly distasteful and that during the wet season they advertise this fact with bright colours. Because insects are abundant, birds and other predators learn to avoid them. However, during the dry season insects are relatively scarce and birds are not so particular. Under these circumstances it might be an advantage for the butterflies to be cryptically coloured.

COLOUR CHANGES ASSOCIATED WITH GROWING UP

Many animals change colour at particular times in their development towards maturity, their different colours reflecting their changing needs. In young animals the need is often for concealment. Young terns, plovers and oystercatchers, for example, are cryptic because they hatch in the open and would seldom survive if they resembled their conspicuous parents.

It is often thought that young wild pigs are striped and spotted for the same reason, but this is probably not the case. For one reason, they do not behave cryptically, their reaction to danger being to bolt with their mother rather than to hide. More significant still, their mother never leaves them and will defend them with the utmost ferocity. It seems more likely that the spots and stripes are a *social* signal which elicits appropriate responses from the mother and other adults. The same may be true of the spots of young lions, though a suggested alternative explanation, which might also be true of the pigs, is that they are an ancestral character which remains because it carries no positive disadvantage.

Colour changes associated with development and also with concealment occur in the marine molluscs known as sea-hares. One species—*Aplysia punctata*—migrates from deep water to the shore as it grows older, and it changes through a succession of colours—from rose-red to red-brown and finally to olive-brown. These colour changes ensure that it remains cryptic throughout its life, for they are correlated exactly with the colour of the succession of seaweeds on which it lives and feeds during its migration. The same colour changes occur even in the laboratory when a constant food supply is provided. Obviously the colour changes of *Aplysia punctata* cannot be caused by the accumulation of pigments from its food; the pigments must be synthesized and the timing of their appearance must be genetically determined. In this respect *Aplysia punctata* and other sea-hares differ markedly from those not-so-distantly related Nudibranch sea-slugs which match their background by accumulating pigments from the sponges upon which they browse.

CAMOUFLAGE AND CONCEALMENT OF SHADOW

Up to this point we have been considering the multitude of different ways in which camouflaged or disguised animals contrive to match the colour of their background. As far as a camouflaged animal is concerned this is only the first step, for no matter how perfectly it matches its background it will still be very conspicuous if its shape and outline is thrown into relief by body shadows, or if it casts a strong shadow on the ground. These visual cues need to be obscured by more sophisticated camouflage techniques, such as behavioural and struc-

To reduce its profile and conspicuous shadow, a fallow deer fawn crouches low and stretches its head and neck along the ground. Its spotted coat breaks up its shape and blends well with the sun-dappled woodland vegetation in which it hides.

ROSS

Plover chicks have flecked or disruptive patterns which blend beautifully with a background of sand or pebbles, but their camouflage is only effective when it is combined with special behaviour (see below). This standing Kentish plover chick, unalarmed, casts a shadow far more conspicuous than itself, particularly when it moves.

Responding to its parents' alarm calls, the Kittlitz's plover chick crouches low and motionless, casting no shadow, and blending perfectly with its sandy background.

The structurally flattened body of the plaice merges so imperceptibly with the seabed that it casts very little visible shadow. Like other flat-fishes, the plaice can change colour to match any background and usually makes itself still less conspicuous by covering its body with mud, sand or gravel.

FOGDEN

WARD

In the diffuse light of its African forest home, this flattened praying mantis is almost indistinguishable from the bark of the parasol tree. However, the electronic flash used for the photograph has produced a sharp shadow, illustrating the way in which shadows can detract from the efficiency of even the most perfect colour resemblance.

tural flattening, countershading and disruptive colouration. We shall first consider concealment of shadows.

The cryptic animals most vulnerable to detection by their shadows are those that live on flat open ground. Racing crabs scurrying across sandy beaches are readily detected by their shadow, but so well do they match their sandy background that their shadow often appears to have no material cause at all. Not for nothing are they often called ghost crabs, and the name is particularly suitable when they are seen by moonlight or by torchlight. When danger threatens, racing crabs often react by crouching. This behavioural flattening obscures their outline and minimizes the size and conspicuousness of their shadow. Crouching in one form or another is found in a great variety of cryptic animals, including antelope and deer, many ground-dwelling birds, lizards, frogs and insects.

Shadows may also be eliminated by the way an animal orientates itself with respect to the sun. Butterflies which close their wings above their back tend to alight facing the sun so that their shadow is reduced to a thin line. Some nightjars have a similar habit and gradually revolve as the sun changes position in the sky. Butterflies which are particularly vulnerable to detection from above tilt themselves to one side so that their shadow is covered by the cryptic underside of their wings. The green hairstreak is a good example and can tilt almost flat against the leaf upon which it is settled.

Many cryptic animals have solved the problem of shadows by structural rather than behavioural means. They have done so by evolving a flattened form which enables them to merge themselves smoothly with the lines of whatever substrate they are on. Some of the most extreme examples are fish, such as plaice and rays, which lie flat on the seabed and cover the edges of their body with mud or sand, so that there is no possibility of even the tiniest shadow. It is interesting to note that plaice and their relatives are flattened laterally, so that strictly speaking they rest on their side, while rays

FOGDEN/COLEMAN

Left: The caterpillar of the Malaysian archduke butterfly obscures its shadow by means of long lateral spines pressed close to the leaf upon which it sits; it also has a pale dorsal stripe which resembles the midrib of the leaf and positions itself appropriately. The horned lizard in the picture on page 39 has spiny lateral scales which obscure shadows in much the same way as the spines of this caterpillar.

Right: The caterpillar of the silkmoth, *Attacus sp.*, which habitually rests upside-down, is shaded darker on its underparts than its back and illustrates the principle of countershading. In natural, diffuse light, it appears evenly coloured, flat and leaf-like; by contrast, in the same amount of light when illuminated from below, its pale back and solid, rounded shape are very conspicuous (below).

and skates are flattened dorso-ventrally and rest on their belly. The two groups of fish have achieved the same end result in different ways. Many animals accentuate their flattened form with lateral flaps and flanges, good examples being found among geckos which live on tree-trunks; *Uroplates fimbriatus* from Madagascar has a frill all around its head and body and lateral flaps on its tail, while the flying gecko—*Ptychozoon kuhli*—from South-east Asia has very broad flaps which it also uses to a limited extent for gliding from tree to tree, or parachuting to the ground to escape predators. There are a whole host of insects which have evolved lateral flanges, including mantids, bugs and beetles, and there are tropical caterpillars which have achieved the same end by means of lateral spines. The most numerous and familiar examples of all are moths, most of which fold their wings laterally and press them against the tree trunk, leaf or stone upon which they are resting.

COUNTERSHADING

Shadows also give away the presence of an animal in another way, which can best be understood by considering the effect of sunlight on an evenly coloured object such as a sphere. Because it is illuminated from above the sphere appears lightest above and gradually shades to a darker colour below. The shading throws the sphere into relief and gives it the appearance of solidity which enables us to distinguish its shape. By using paint to shade the sphere darker above and paler below it is possible to exactly counterbalance the natural shading. If done perfectly the solid appearance of

the sphere disappears and it looks like a uniformly coloured thin flat disc. This process of optical flattening is known as countershading.

Countershading is used to destroy the appearance of solidity in animals. It is almost universal among cryptic species and greatly increases the effectiveness of colour resemblance. Countershading is usually achieved by the simple gradation of colour that can be seen in the lion, many antelope, deer, pelagic fish and a host of other animals, but it can also be achieved by the use of bold patterns of spots and stripes which blend when viewed from a distance. The zebra, for example, has dark stripes which are widest on its back; the cheetah has dark spots which are more numerous on its back than its belly; and the guineafowl, which has light-coloured spots on a dark background, has smaller spots on its back than its breast. When viewed at a distance all three of these patterns blend to produce the same effect as countershading.

There are many interesting variations to the theme of countershading. For example, the Nile catfish habitually swims upside-down and is therefore darker on its belly than its back. The same reversal of countershading can be seen on many caterpillars, particularly those of hawkmoths and silkmoths, which hang upside-down on their food plants. By way of contrast, the caterpillar of the purple emperor butterfly rests in a perpendicular position with its head uppermost, and is shaded darkest on its head and palest behind. The cuttlefish, supreme master of rapid colour change, adjusts its shading to fit its orientation, whichever part of its body is uppermost being instantly shaded darker than the rest. Particularly intriguing are a number

46

47

of transparent fishes, such as the coral fish, *Coralliozetus cardonae*, which have countershaded internal organs visible through their transparent skin and body muscles. Examples such as these emphasize the adaptiveness of countershading and its great importance to camouflage.

DISRUPTIVE COLOURATION

Colour resemblance, shadow elimination and countershading go a long way towards obscuring an animal, but unless the colour match is exact and the lighting perfect (neither of which will often be the case in real life) the animal is still likely to be visible as a continuous patch of colour with a recognizable outline. It is important that this continuity of surface and outline should be broken up, and this is most commonly achieved by means of disruptive markings in the form of irregularly shaped patches of contrasting colours. The individual patches are more conspicuous than the animal itself, but their shapes bear no obvious relation to the animal's own shape and so give no indication of its existence. It is an advantage if some of the patches match the colour of parts of the immediate surroundings. This contributes to the

Though conspicuous at close quarters, the black and white stripes of the zebra are just another way of achieving countershading; they are wider above than below and blend to a uniform grey when seen at a distance. In the dry season, when savannah vegetation is dry and pale, the stripes probably provide an effective camouflage, especially at dawn and dusk, when the zebra is most vulnerable to its major predator—the lion.

This forest toad from Borneo has a disruptive pattern which merges perfectly with the litter of dead leaves on the forest floor. Its pale, dorsal stripe resembles the midrib or vein of a leaf and disrupts its bilateral symmetry.

The smooth surface of the African gaboon viper is transformed by its strong disruptive pattern into an apparently haphazard collection of black, brown, buff and grey leaf-like surfaces. Triangular black patches divert attention from its eyes and appear to be shadows cast by the flat leaf-like surface of its head.

disruptive effect by causing some but not all of the animal's outline to merge with its background. This effect is known as differential blending.

Disruptive colouration transforms smooth surfaces into apparently haphazard collections of irregular or even leaf-like surfaces. The gaboon viper is a wonderful example of this effect. Viewed against a plain background it would be most conspicuous, but its contrasting pattern of fawn, brown and black merges exactly with the pattern made by leaves, light and shade on the forest floor where it lives.

Disruptive colouration also breaks up the shape of an animal into two or more apparently unrelated parts. This may be accomplished by means of a conspicuous stripe along the middle of the animal's back, as in many frogs, toads, moths and grass-

hoppers. This has the important effect of disrupting its bilateral symmetry, because the shape of half an animal is asymmetrical and therefore less familiar and less easily recognizable than the animal itself. Most naturalists will have discovered for themselves that the symmetry of an animal often provides the cue which leads to its camouflage being penetrated, and will readily appreciate that any adaptation that helps to obscure symmetry is of great benefit.

In addition to breaking up continuous surfaces on the body of an animal, disruptive colouration can also be used to make different parts of the body appear joined together. This joining effect obscures the characteristic shapes of the separate parts and is accomplished by what is known as coincident disruptive colouration. It is well illustrated by many

49

frogs and insects. For example, most cryptic moths that rest on tree-trunks have a bold disruptive pattern that is continuous across their body and both pairs of wings. Similarly cryptic butterflies which rest among dead leaves have a disruptive pattern which continues apparently uninterrupted from the forewing to the hindwing. In each case, the pattern cuts across the existing structures of the animal, and superimposes the appearance of a new structure which resembles the background rather than the animal or any of its parts. It has to be emphasized that coincident disruptive colouration is only effective when an animal is in its natural resting position. Traditionally set in completely un-natural positions, moths and butterflies in a collection do not show this continuity of pattern.

CONCEALMENT OF THE EYE

Concentric circles in contrasting colours are probably more conspicuous than any other pattern, which is why such a pattern is used as a target in archery and rifle shooting. The eye of an animal is conspicuous for the same reason, especially as it also glistens. It presents a camouflage problem which has been solved in a variety of ways.

In the case of many nocturnal snakes, lizards and frogs, which are cryptic by day, the eye is obscured by reducing the shape of the pupil to a tiny vertical or horizontal slit, and by matching the colour of the iris to that of the skin surrounding the eye. A more remarkable example is that of the fish, *Petrometopon cruentatus*, from Tortugas, in which the conjunctiva of the eye, as well as the surrounding skin, is covered with dark spots among which the pupil is hardly

The pine hawkmoth (left) and waved umber moth are examples of background picturing, having patterns which picture the bark of the trees upon which they rest. The pine hawkmoth sits vertically, the waved umber moth horizontally; in any other position they would be more conspicuous for their crevice-like markings would be unaligned with the genuine crevices in the bark.

When alarmed, the bittern stretches its head, neck and body vertically, so that the streaks in its plumage run parallel to the surrounding reeds.

noticeable. What is remarkable about this case is that the spots extend to parts of the conjunctiva that are normally hidden. As a result, the pupil remains inconspicuous even when the eye is moved.

In the case of many other animals the eyelids rather than the iris match the colour of the surrounding body. When danger threatens the eye is closed to the merest slit and it becomes almost invisible. Superb examples of this technique are found among nocturnal birds, such as nightjars and frogmouths, and in the Australian spiny lizard. In the case of chameleons only the pupil of the eye is ever visible, because the rest of the eye is permanently covered by skin which matches the surrounding body. This extreme adaptation probably results from the fact that chameleons are diurnal feeders and are unable to sit around for long with

their eyes shut. Though insects lack eyelids at least one cryptic species—the herald moth—has evolved tufts of hairs which suffice just as well to conceal its eyes. When the moth takes flight the tufts are pulled aside by its antennae swinging forward.

The commonest method of camouflaging the eye is probably by means of disruptive patches or stripes which enclose it or break up its outline, thereby masking its shape. There are a host of examples among snakes, frogs, fish and many insects, especially grasshoppers. The dark eye-stripes of cryptic birds, such as snipe and woodcock, probably have the same function as well, but the conspicuous eye-stripes and patches on many other species, nuthatches and tits for example, are more likely to have a social function. Generally, patterns that function as eye camouflage are part of an overall disruptive pattern that camouflages the whole animal, though there are a few exceptional cases where genuine disruptive eye-stripes are found in animals which are otherwise brightly coloured and conspicuous. For example, many brilliantly coloured butterfly fishes, *Chaetodon*, camouflage their eyes to protect them from sabre-toothed blennies. These latter fish feed by biting pieces of skin from larger fish and they attack their eyes preferentially.

BACKGROUND PICTURING

Just as the colours applied by an artist may create a picture which bears no resemblance to the shape and texture of his canvas, so simple disruptive colouration creates the illusion of haphazard shapes and textures which bear no relation to the structure of the animal. From this illusion of haphazard shapes and textures it is only a short step to a camouflage technique known as background picturing, in which the shapes and textures of an animal's surroundings are reproduced in its colouring. This effect can be seen in the striped patterns of such grass or reed haunting birds as bitterns, quails, pipits and grass-warblers, *Cisticola*, among many others, as well as in innumerable frogs and insects. It is seen again in the spotted and stippled patterns of animals that live in stony or sandy deserts, while animals that live on tree-trunks often have patterns that depict the cracks and crevices of bark, or the scales and filaments of lichen. Similar effects can be seen in animals belonging to every environment from mountain tops to shallow seas. Not infrequently the patterns involved would be conspicuous if seen out of context. The black and

The assymmetrical pattern of the East African frog, *Chiromantis*, accurately pictures the rough surface of the tree-bark upon which it lives.

orange stripes of the tiger, for example, are obvious enough in the bleak context of a zoo cage, but they match the streaky pattern of brilliant light and deep shade cast by a tropical sun in the long, dry grass and reeds in which the tiger lives for most of the year. An equally gaudy black, brown and orange bat, *Kerivoula*, from Taiwan pictures the decaying leaves of the longan trees in which it roosts. Examples such as these emphasize the importance of seeing an animal in its natural surroundings before speculating on the function of its colouring.

The fact that natural backgrounds usually have a definite pattern makes appropriate orientation even more important in background picturing than in simple disruptive colouration. For example, the cracks and crevices in the bark of a tree run along the trunk, and the markings of a cryptic animal must do the same. It is for this reason that bitterns stretch upright when danger threatens, because only then does their streaky plumage merge perfectly with the vertical pattern of reeds among which they live. Moreover, bitterns sway in unison with surrounding reeds being blown by the wind.

DISGUISE

Just as background picturing is a step in advance of other camouflage techniques, so disguise is a step in advance of background picturing. Instead of giving an animal the appearance of a 'generalized piece' of its environment—the appearance of a

Grasshoppers of many kinds mimic leaves. This longhorn grass-hopper, or katydid, from the forests of Borneo, mimics the shiny, dark green leaves of a forest tree, matching them in its colour, texture and simulated leaf veins.

Right: The South-east Asian walking leaf, *Phyllium*, is the most famous of all leaf mimics. Here it is photographed in captivity, but its disguise is even more effective in its natural surroundings. Males are rare, less leaf-like than females and have functional wings.

The shorthorn grasshopper, *Eumegaladon*, from Borneo, mimics a half-eaten, dead, brown leaf on the forest floor. Its elongated head and antennae resemble a leaf-stalk and, like many leaf mimics, it has realistic imitation leaf veins and mould spots. When danger threatens, it tilts to one side, to look like a leaf lying flat on the ground. At night it climbs into the forest undergrowth to feed on fresh green leaves.

patch of reeds in the case of a bittern—disguise gives it the appearance of a specific inanimate object in the environment—a leaf, twig, stone or even a bird-dropping. Disguised animals generally have extensive structural modifications, but have no need of such camouflage techniques as shadow concealment and countershading; an animal disguised as a leaf or twig simply casts its shadow in the same way as all the other leaves and twigs around it. Nevertheless, disguise is not a sharply definable category. For example, the two geckos—*Uroplates fimbriatus* and *Ptychozoon kuhli*—which were discussed with respect to camouflage, might just as appropriately have been classified as animals disguised as bark.

Disguise can be regarded as a type of mimicry, the main models being leaves, twigs and small stones, though a host of other objects are used less often. The intricate perfection of some examples of disguise is difficult to appreciate from photographs. Still less is it possible to appreciate how effective the disguises are in the wild, for by framing an animal, a photograph makes it relatively easy to see. It is a very different matter in the great tropical rain forests of Malaysia, Africa and South America, where examples of disguise reach their greatest

perfection and sophistication. Finding leaf-insects and stick-insects among the lush vegetation is almost as much a matter of luck as skill. Certainly the birds find it difficult too. A banded broadbill was once seen to catch a leaf-insect which had been brushed from a sapling by a passing deer. It then spent several minutes apparently searching for more by pecking at leaves at random—without success.

LEAF MIMICRY

Many animals mimic leaves. The most successful of all are insects, whose veined wings are ideally suited for expansion and modification into leaf-like shapes and surfaces, but there are also excellent vertebrate examples among chameleons, toads, frogs and fish. All sorts of leaves are used as models —fresh green leaves, dead brown leaves, clusters of

The South American glass-winged butterfly, *Pteronymia*, blends with the surrounding vegetation by having transparent wings through which the vegetation is visible.

leaves, fragments of leaves, rotting leaves and even falling leaves.

Perhaps the most perfect of all leaf mimics is the so-called walking leaf, *Phyllium*, from Malaysia and Indonesia. It has curious flat flanges on its thorax and legs, an abdomen and wings that are paper thin and marked with leaf-like venations, and edges to its body that look chewed or torn. In fact, it perfectly resembles a cluster of green leaves and leaf fragments, which is quite appropriate as it is herbivorous itself and lives among half-eaten leaves. A walking leaf's disguise is enhanced by its behaviour. It moves very slowly, advancing one leg at a time, and it sways to and fro in imitation of leaves in a gentle breeze. Perhaps the ultimate testimony to the perfection of the disguise of *Phyllium* is that other leaf-eating insects have been known to take a bite at it.

Apart from *Phyllium* the best mimics of green leaves are found among praying mantids and grasshoppers, the praying mantids generally being flattened horizontally like *Phyllium*, whereas the grasshoppers are more often compressed from side to side. Examples of both groups have marvellously realistic 'mould spots', 'insect-chewed' edges and 'diseased' or 'rotted' patches. In fact they are so realistic that one almost believes them genuine until identical marks are seen reproduced in the exact same spot on every specimen. Many praying mantids and grasshoppers mimic dead leaves equally well. The green forms and the brown forms always segregate appropriately onto green leaves and dead leaves respectively, though the brown forms of grasshoppers move to green leaves to feed at night.

The dead leaf butterfly, *Kallima*, is another classic of leaf mimicry. It is a brightly coloured butterfly on its upper surface, with beautiful iridescent markings, but in its natural resting

Like numerous butterflies and moths, the South African owlet moth is disguised as a dead leaf. In its natural resting position it has a leaf-like shape and a simulated midrib which continues across both pairs of wings.

FLETCHER/NSP

position, with its wings folded above its back, it is an exact replica of a dead leaf. It has a false midrib and other leaf-like veins running across both fore and hindwings, a pair of short tails which mimic a leaf-stalk, and it often has 'mould spots' and 'holes' apparently cut by leaf-cutting insects. Its behaviour is equally perfect, for it always settles in a leaf-like posture and usually upon a branch bearing at least some dead leaves. The great naturalist and evolutionist Alfred Wallace described how *Kallima* disappeared as if by magic when it settled, and how he had difficulty in recognizing a specimen even after he had marked the exact spot where it alighted. Many other butterflies resemble leaves in one way or another. Some species have transparent windows in their wings which resemble partly skeletonized leaf patches, while the wings of glass-winged butterflies are transparent so that leaves are visible through them. Some butterflies even contrive to resemble a leaf while flying, having developed a hesitant, fluttering flight that mimics the movements of a falling leaf.

Leaf-mimicking moths are even more abundant than butterflies, and they usually differ in exposing the upper rather than the lower surfaces of their wings. In many species the leaf pattern extends over both forewings and hindwings, but in others, including many hawkmoths and silkmoths, the hindwings have brilliant markings which are normally covered. As in butterflies, there are some species that mimic skeletonized leaves, a superb example being *Draconia rusina* from South America, and others that have transparent wings through which real leaves are visible. Other moths, such as the angle-shades, resemble crumpled or curled leaves. This is a useful form of leaf disguise which does away with the need for flattening, and one which is quite frequent among caterpillars and pupae of both butterflies and moths. Other caterpillars, including such large fat examples as those of hawkmoths and silkmoths, manage to mimic flat leaves by a combination of countershading and leaf-like markings. The countershading gives their rounded form the appearance of flatness, while leaf veins are often imitated with diagonal stripes, particularly among the hawkmoths.

A variation on the theme of leaf disguise is found among many stick-insects and spiders which mimic not the blade of a leaf, but its midrib and veins. All the species concerned have long attenuated bodies which they orientate along the real midrib of a leaf.

This East African silk moth caterpillar is successfully disguised as an acacia leaf.

Left: The leaf disguise of the lappet moth is enhanced by the scalloped edges of its wings and its habit of disrupting its moth-like shape by exposing its hindwings in front of its forewings.

Their legs are held in pairs backwards and forwards as continuations of the body, or may be held sideways to resemble veins branching from the midrib. It is essential that the animals orientate themselves accurately for the disguise to be effective.

The most remarkable leaf mimics among the vertebrates are undoubtedly the South American leaf fish from the Amazon valley, and various species of bat fish from Indo-Pacific coasts. The leaf fish is a generalized leaf mimic which lives in fresh-water streams, while the bat fish specifically resembles mangrove leaves and lives in the shallow brackish water in which mangroves grow. Both the leaf fish and the bat fish have a strongly compressed body with a suggestion of leaf venations and mould markings, but are otherwise quite different in the way they are structurally modified to resemble a leaf. In the case of the leaf fish the leaf-stalk and leaf-tip are depicted by a short barbel on the tip of the lower jaw and a tapering pigmented tail fin. By contrast, the bat fish's leaf-stalk and leaf-tip are depicted by its elongated and pigmented dorsal and ventral fins, while its tail fin, which would otherwise spoil the line of the 'leaf', is transparent. In other words, the construction of the leaf fish resembles that of a horizontal leaf, that of the bat fish a vertical leaf. The leaf fish is a predator of other fish and makes use of its disguise to stalk its prey. It floats on its side, or head-down, leaf-like, just below the surface of the water, and propels itself along with no apparent movement using only its transparent dorsal and anal fins. In this way, looking like a drifting, water-logged leaf, it stealthily approaches unsuspecting fish, until with a final lunge it engulfs its prey. On the other hand, the disguise of the bat fish seems to be mainly protective. It floats gently along, moving only its transparent tail, sometimes twisting or swaying in exactly the manner of drifting mangrove leaves. When danger threatens it topples inertly to the sea bottom where it lies quietly among the leaves it so much resembles.

Among other vertebrate leaf mimics a number of toads and frogs are particularly interesting for the way in which they combine structural modifications

PAYSAN

The South American leaf fish lives among submerged and drifting leaves in the clear, still water of forest streams in the Amazon basin. Its leaf-like markings are very variable and it is able to adjust its colour to match changing conditions. When stalking prey it appears to drift through the water like a water-logged leaf.

In the shallow coastal waters of the Indo-Pacific region, usually in the vicinity of mangroves, there are several species of bat fish which resemble mangrove leaves. They swim like drifting leaves or, when alarmed, lie inertly on their side on the seabed.

WARD

with shading effects. Toads and frogs are always relatively portly creatures but such species as *Megophrys nasuta* of Malaysia, *Bufo typhonius* of South America, and *B. superciliaris* of Africa show a considerable degree of structural flattening. All three species emphasize their flattened appearance by means of shading. Each has a pale and peculiarly flattened back sharply edged on either side by flanges of skin, while below the flanges the sides of the head and body are usually black. As a result, each species looks exactly like a flat pale leaf casting a deep shadow. Similar effects can be seen on several other toads and frogs, and on the head of the gaboon viper and other snakes. In all these examples the cryptic effect is achieved by shading that is the reverse of countershading. The one type produces an illusion of thinness in a horizontal plane, the other in a vertical plane.

STICK MIMICRY

Numerous insects and a few vertebrates have adopted stick- and twig-like disguises, the most specialized and perfect examples being among Geometrid (looper) caterpillars and Phasmid insects. Stick-caterpillars resemble small twigs growing on the trees on which they feed, the resemblance often being perfect even to the tiniest detail of structure and texture. As already mentioned, there are several species which are polymorphic, each form mimicking a growing twig from a different food plant. The posture of stick-caterpillars greatly enhances their disguise, for they grasp a stem with their hind claspers, and incline their body outwards resembling a branching stem. Some species remain

FOGDEN

The Malaysian horned frog lives on the forest floor and is one of the most remarkable of leaf mimics. Projecting horns of skin, shaded black below, effectively conceal its eyes.

Below, left: Disguised as a shrivelled yellow leaf, this spider from Borneo sits on its web among the leaves that conceal its egg cocoons. Its legs are carefully arranged to resemble the pointed tip of the leaf and its extraordinarily elongated abdomen ends in a petiole-like structure.

Like a number of spiders, caterpillars, mantids and other stick insects, this stick insect from the forests of Borneo positions itself on a leaf so as to look like the leaf's midrib and veins.

The looper caterpillars of numerous Geometrid moths are superbly disguised as twigs of the trees on which they feed. As a second line of defence, they can drop from their branch, suspending themselves on a silken thread.

The twig-like disguise of this Malaysian praying mantis conceals it from predators and prey alike.

Below, right: This small moth mimics a bird-dropping and is difficult to distinguish from the genuine droppings among which it is sitting.

Like most herbivorous insects disguised as dead leaves or broken twigs, the thorny stick-insect, *Dares nolimetangere*, from Borneo, feeds only at night and spends the day in the litter of leaves and twigs on the forest floor.

in this attitude throughout the day and feed at night, while others assume it only when danger threatens. The attitude demands considerable muscular control, but some species ease the strain by partially supporting their head and body with a silk thread which is anchored above them.

Many Phasmid stick-insects disguise themselves as growing twigs in much the same way as Geometrid caterpillars, but rest head-down with their abdomen, rather than their head, inclined at an angle to the stem. Though achieved differently, the effect resembles that of stick-caterpillars and is almost equally perfect. In another variation on this theme an Australian twig-mimicking grasshopper achieves the same effect by inclining its pigmented forewings in imitation of a lateral twig, the rest of its body and all its legs being flattened against a stem. Other Phasmid stick-insects, including many inhabiting forests, mimic dead or broken twigs and spend the day in the litter of leaves and twigs which covers the ground. At night they climb into the undergrowth to feed, but even then they are difficult to find, because they relax their hold and fall to the ground at the slightest disturbance. This type of stick-insect includes some that are among the largest insects in the world, reaching well over a foot in length and an inch or more in width.

Other insects which mimic twigs include praying mantids and moths. The latter, because of their stumpy shape, are only able to mimic short broken stumps of twigs. Notable examples are the buff-tipped moth from Europe and *Duomitus leuconotus* from India. The latter has spiky growths on its head resembling a splintered twig and markings on its wings resembling patches of lichens.

Resembling a thin, green vine or liana, the Malaysian grass-green whip-snake lives in the branches of trees and bushes, where it ambushes small birds and arboreal lizards.

Several African species of Flattid bugs arrange themselves on plant stems so as to resemble a spike of flowers.

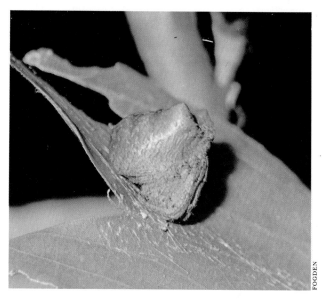

This Malaysian spider catches prey only at night. Every dusk it spins a new web and every dawn it dismantles and eats it. During the day it sits on a leaf, exposed and immobile, resembling a curled brown leaf or plant gall.

Right: These two specimens of an East African moth resemble patches of lichen. In this species no two individuals are alike, making it particularly difficult for predators to form a strong searching image for them.

The only vertebrates which can be said to be disguised as sticks or twigs are a number of snakes, though there are several birds (discussed later) that are disguised as snags or stumps of wood. One of the most realistic stick-mimicking snakes must certainly be the blunt-headed tree-snake of Malaysia. This species has the peculiar habit of lying in a stiff, stick-like posture upon branches of the forest undergrowth or on the upper surface of a palm frond, just as if it were a stick which had fallen from above. Its resemblance to a stick is further enhanced by the shape and colouring of its blunt head which looks like the broken end of a small branch, and by its being less prone than most snakes to flee at the approach of danger. Its disguise almost certainly serves as much to protect it from hornbills (which eat a lot of snakes) and other predators, as to help it ambush prey. Equally specialized are several tree-snakes which resemble vines and lianas rather than sticks. Such snakes are represented by *Dryophis* in South-east Asia, *Oxybelis* in South and Central America, and *Thelotornis* in Africa. All are extremely long and thin, a specimen of *Oxybelis acuminatus* measured by Cott being four feet long and only a quarter of an inch in diameter—or 160 times as long as its width. Species of *Dryophis*, aptly named whip-snakes, and *Oxybelis* are very variable in colour, some of them matching green vines, others the browns and greys of lianas.

DISGUISE AS FLOWERS

Among the most beautiful insects are a few praying mantids and Flattid bugs which disguise themselves as flowers. The praying mantids, notably *Hymenopus coronatus* of South-east Asia and *Idolum diabolicum* of Africa, are remarkable in that their disguise as flowers is alluring as well as protective. The former is discussed in the chapter on alluring colouration. The Flattid bugs are perhaps even more curious and remarkable. Individually they have a rather moth-like appearance and several species are bright green, yellow, pink or white. Sitting on the stem of a plant they look a little like a leguminous flower, such as a broom, especially when several gather on a single plant. In East Africa individuals of species in the genus *Ityraea* carry the flower mimicry a stage further by congregating on a single stem and arranging themselves in the form of a complex lupine-like inflorescence. More remarkable still, *I. nigrocincta* has two colour forms—green and yellow—which

segregate, so it is said, the green individuals gathering at the top of the inflorescence, the yellow at the bottom. The result is a beautiful impression of an inflorescence with yellow flowers and green buds, progressively opening from below in the same way as lupins, foxgloves and many other flowers. Though it is easy to see Flattid bugs congregated on plants, in other parts of the world as well as Africa, it is difficult to find a true 'inflorescence'. They seem to be rather rare.

This example of disguise is curious in that it is rare for cryptic animals to congregate in this way. In fact, it is usually greatly to their advantage to be well-spaced, as already mentioned. On the other hand, congregations are typical of warningly-coloured animals, and indeed typical of several species of Flattid bugs which are waxy and distasteful. It is not known whether or not *Ityraea* species are distasteful to predators, but the detailed resemblance of a congregation to an inflorescence certainly suggests that their true identity must be disguised to potential predators, such as birds. If so, how did the congregating habit evolve in a cryptic species? A possible answer is that the 'ancestral inflorescence' of *Ityraea* was a vaguely flower-like congregation of distasteful, white, waxy Flattids. If their resemblance to a flower eventually proved a greater advantage than their being waxy and distasteful, there would have been natural selection to enhance their flower resemblance still further, possibly culminating in a complex 'inflorescence' like that of *Ityraea nigrocincta*.

DISGUISE AS BIRD-DROPPINGS

Mimicry of bird-droppings is probably the most bizarre of the many disguises adopted by animals. It is fairly common among moths and caterpillars in which its function is solely protective. Most of these moths resemble flattened splashes of excrement dropped by flying birds from a considerable height, while caterpillars more often resemble rounded, half-dried blobs of excrement dropped from a lesser height. To convey the extraordinary likeness of some bird-dropping mimics to the real thing one can do no better than quote the experience of Colonel A. Newnham, in India, as recorded in Poulton's book:
'I was stretching across to collect a beetle and in withdrawing my hand nearly touched what I took to be the disgusting excrement of a crow. Then to my astonishment I saw it was a caterpillar half-hanging, half-lying limply down a leaf. The chief

thing that struck me about it was the unnecessary perfection of the resemblance. The Russian proverb "Nature dura" (Nature is a fool) occurred to me, because it seems so much simpler and efficacious for the larva when at rest to conceal itself under the leaf. Another thing that struck me was the skill with which the colouring rendered the varying surfaces, the dried portion at the top, then the main portion, moist, viscid, soft, and the glistening globule at the end. A skilled artist, working with all the materials at his command, could not have done it better.'
The caterpillar concerned was of a Notodontid moth. Another interesting example is provided by

This African spider is disguised as thorns on an acacia branch. Like the gall-mimicking Malaysian species, to which it is closely related, it catches prey only at night, spinning a new web each day.

the young caterpillars of an African Bombycid moth, *Triloqua obliquissima*, which crowd together on the surface of a leaf and jointly resemble a patch of excrement. Perhaps the most ingenious of all the bird-dropping mimics is a spider, discussed later, in which the disguise is alluring as well as protective.

DISGUISE AS OTHER OBJECTS

One could add almost indefinitely to the list of objects that are imitated by terrestrial animals. No mention has been made of thorns, stipules, seeds, plant galls, lichens or stones, and doubtless many

new deceptions remain to be discovered, especially in the tropics. One of the best disguises we ever came across was a spider in Sarawak which looked like an indeterminate piece of vegetation—perhaps a nodule of bark that had fallen from a tree above, a curled leaf, or even a plant gall. Although it was not at all clear what it was disguised as, it certainly did not look like a spider. This species spends the day crouched in full view on the surface of a leaf, its legs pressed tightly against its body. At dusk it becomes active and sets to work to spin its web, upon which it sits throughout the night. At dawn it wraps up its web, eats it, and resumes its cryptic

Covered in strange protuberances and filamentous growths, the sargassum fish is amazingly difficult to see among the masses of floating weed in the Sargasso Sea.

Masking crabs disguise themselves with pieces of seaweed which they attach to hooked bristles on their carapace and legs.

The larvae of Chrysomelid lily beetles feed on the flowers of lilies and orchids and disguise themselves as insect-droppings by coating themselves with their own excrement.

At the least sign of danger, the conspicuous oystercatcher keeps well away from its eggs or young, which are well protected by their excellent camouflage.

position on its leaf. The success of this spider's disguise is dependent on its posture and particularly the concealment of its legs. It is interesting that the segments of its legs which are totally concealed by day are dark red and black, in contrast to the colour of the visible parts of its body.

DISGUISE IN MARINE ANIMALS

Among marine animals the numerous small fishes, crabs, shrimps and molluscs that match sessile animals, such as corals, sponges and tunicates, are probably best classified as cases of background picturing. The sea-slugs mentioned in the section on colour resemblance related to diet are good examples. However, there is a great diversity of animals, from fishes down to Annelid worms, that are disguised as seaweed in the most minute detail.

Examples are especially numerous among the enormous masses of weed that float in the Sargasso Sea, the most striking being the bizarre sargassum fishes, such as *Histrio histrio* and *Antennarius marmoratus*, which are adorned with peculiar weed-like filaments, resembling those of sargassum weed. Even more bizarre is the sea-dragon, an Australian sea-horse whose fantastic outline is profusely covered with broad fronds like those of the seaweed in which it lives. Other weed-mimicking fishes occur in beds of eel-grass, and are interesting for the way in which they imitate the upright blades of eel-grass by adopting a vertical head-up or head-down posture. Various pipe-fishes, with their slender sinuous bodies, are particularly well disguised and usually adopt a head-up posture, while the slender trigger-fish assumes a head-down posture.

66

ADVENTITIOUS DISGUISE

Adventitious disguise is 'acquired' disguise, and describes the habit that some animals have of covering themselves with soil, leaf fragments, bark, seaweed or other debris from their surroundings. Adventitious disguise not only produces a literally perfect resemblance to particular surroundings, but also gives a greater freedom of choice of surroundings to species which are neither polymorphic nor able to change colour. Caddis-fly larvae, for example, build themselves portable tubular cases with sand-grains when they live in fast-flowing streams, and with tiny pieces of wood and leaves when they live in sluggish streams or ponds. Many other insects make use of adventitious disguises: the caterpillars of bag-worm moths cover themselves with pieces of leaves and twigs in much the same way as caddis-fly larvae, and various beetles cover themselves with sand, mud or whatever other material is available. In Borneo we came across two splendid examples: a Geometrid caterpillar which covers itself with flower-buds attached to spines on its body; and, more curious still, a Chrysomelid beetle, whose larvae feed on the flowers of orchids, covering themselves with, and so resembling, blobs of their own excrement.

Among marine animals the masking crabs—*Hyas, Maia, Inachus, Stenorhynchus* and others—have superb adventitious disguises of seaweed which they change according to their different surroundings. The way in which species of *Inachus* and *Stenorhynchus* painstakingly 'dress' themselves has been admirably described by W. Bateson:

'The crab takes a piece of weed in his two chelae, and neither snatching nor biting it, deliberately tears it across as a man tears paper with his hands. He then puts one end of it into his mouth, and, after chewing it up, presumably to soften it, takes it out in the chelae and rubs it firmly on his head or legs until it is caught by the peculiar curved hairs which cover them. If the piece of weed is not caught by the hairs, the crab puts it back in his mouth and chews it up again. The whole proceeding is most human and purposeful. Many substances such as hydroids, sponges, Polyzoa, and weeds of many kinds and colours are thus used, but these various substances are nearly always symmetrically placed on corresponding parts of the body, and particularly long plume-like pieces are fixed on the head, sticking up from it.'

Other crabs are even more specialized in their use of materials to disguise themselves. The sponge

FIEDLER/COLEMAN

crab, for example, uses only a carefully cut piece of sponge, which it holds above its back with specially adapted hindlegs used for no other purpose, and *Dorippe astuta*, which lives in shallow estuarine waters in Borneo, holds a leaf above its back so as to be completely invisible from above.

However, not all adventitious disguises are portable. Some spiders, for example, strew their webs with prey remains, leaves and bits of bark, among which their own body is relatively inconspicuous. Others attach a rolled-leaf shelter to their webs, from which they emerge only to tackle their prey. Similarly, the predatory larva of a South American Neuropteron fly ambushes its prey from the cover of a mound of moss or lichens. Examples such as these gradually intergrade with more conventional shelters, such as holes in the ground or trees, which are outside the scope of disguise, adventitious or otherwise.

CAMOUFLAGE AND DISGUISE IN BIRDS

Though relatively few birds escape predators by being cryptic, those that do so are some of the most perfect of all examples of camouflage and disguise. They merit separate consideration because they provide a useful summary of the principles involved, as well as many fascinating examples of the ways in which cryptic colouring and behaviour interact.

Most camouflaged birds are species, such as ptarmigan, pheasants, bustards, plovers and nightjars, which live and nest on the ground in open country. In many it is only the eggs and young that are camouflaged, but in others the adults too are

Left: Like the oystercatcher, the ringed plover leaves its superbly camouflaged eggs as soon as danger threatens. In common with other ground-nesting birds with cryptic eggs, both species remove egg-shell from their nest as soon as their young hatch, because the insides of the empty shells are conspicuous to predators.

Unlike the oystercatcher and ringed plover, the cryptic female pheasant sits tight on her eggs when predators approach, for the eggs are conspicuous if left uncovered. The gaudy male pheasant takes no part in incubating the eggs.

HOSKING

The female willow ptarmigan is cryptic and incubates her eggs with no active help from the male. The male (right) may help indirectly, however, for he delays moulting into his cryptic summer plumage until after the eggs have hatched and, by being conspicious, may divert the attention of predators from the female.

cryptic, often throughout the year. Of course, it is an advantage for most birds to be inconspicuous by one means or another while breeding, because their eggs and young are very vulnerable for several weeks; however, species that live in wooded country are not visible from a distance, and can generally manage by concealing their nest in dense vegetation and having dull rather than cryptic plumage, while the most brilliantly coloured species, such as parrots, kingfishers and bee-eaters, usually nest completely out of sight in holes.

Birds that nest on the ground can be divided into two distinctly different groups according to how they have solved the problem of concealing their eggs. The species in the first group, of which coursers, plovers and the oystercatcher are good examples, rely on the eggs themselves being cryptic. They are relatively conspicuous species (very, in the case of the oystercatcher) but wary, and they

avoid drawing attention to their eggs by leaving them while potential predators are still far away. The eggs are so well camouflaged that they are difficult to find, even if their position has been carefully marked from a distance, by watching one of the adults return to the nest. The eggs are superb examples of disruptive camouflage and background picturing, often varying in general colour to match the dominant colours of the ground where they are laid. The yellow-wattled plover, for example, lays reddish eggs in an area of brick-red laterite on the Malabar coast of India, but predominantly brown eggs in the dark-soiled surrounding country. Two African species—Temminck's courser and four-banded sandgrouse—have eggs that would be exceptionally conspicuous in any other than the correct surroundings. The courser often has eggs that are almost black but lays them among the scattered, rounded, black

droppings of antelopes, while the sandgrouse has pink eggs and lays them among the pink, fallen leaves of the camel's foot tree, *Bauhinia*. The disadvantage of this method of protecting the eggs is that it is useless if the adults become reluctant to leave their eggs. This does sometimes happen as we know from personal experience in East Africa where we found it difficult to find the nests of crowned plovers and spur-winged plovers in the early morning or evening, but easy during the hottest part of the day when there was a risk of uncovered eggs becoming over-heated. When it was hot the conspicuous adults sat tight and reluctantly left their eggs only when we were so close that there was no chance of our failing to find them, even though they were cryptic.

In the case of species in the second group of ground-nesting birds, which includes many ducks, ptarmigan, pheasants, woodcock and nightjars, at least one of the adults is cryptic and conceals the eggs by sitting on them so tightly that it flushes from them in the face of a predator only at the last possible moment. Once uncovered the eggs themselves tend to be conspicuous, especially in the case of ducks, pheasants and partridges, but normally this never happens. In the ducks, pheasants and many other species in this group only the female is cryptic, and so she alone is able to incubate. The brightly coloured, conspicuous males have little or nothing to do with the rearing of the young, though they probably help indirectly by drawing the attention of predators away from the incubating females. The ptarmigan is particularly interesting, for though the male is cryptic for eight or nine months of the year, he becomes more conspicuous than the incubating female by retaining his white winter plumage until well into the summer when the young have hatched. In other species in this group, such as the partridge, woodcock and nightjars, both sexes are cryptic and share the task of rearing the young. Another example of the way in which colouring and nesting behaviour are inter-related is provided by the sheld-duck, which is an exception among European ducks in that both the male and female are brightly coloured and both incubate the eggs. This is only possible because sheld-ducks nest in rabbit burrows, where their bright colours are hidden and cryptic plumage is no advantage.

Both groups of ground-nesting birds have beautifully cryptic young, covered in down and able to run soon after hatching (though young nightjars differ in shuffling about rather than running, and usually stay close to the nest-site for several days). All the young respond to parental alarm calls by crouching flat on the ground and remaining immobile. In fact, they illustrate all the principles of camouflage: they flatten themselves to avoid casting a shadow, tone with their background, have disruptive patterns, and conceal their eyes by closing them to the merest slits when danger threatens. They can be extraordinarily difficult to find even when one knows they are close by. Cott described how he found young woodcock literally as easy to find by touch as by sight.

Nightjars are probably among the most well-camouflaged of ground-nesting birds. In the Queen Elizabeth Park, in Uganda, two species—the gabon nightjar and the fiery-necked nightjar—breed on bare patches of overgrazed ground, where they are indistinguishable from the scattered pieces

By day, the nocturnal potoo perches in an upright position in which it resembles a stump of wood. When alarmed, it stretches up its body, raises its head and closes its eyes to narrow slits, becoming still more like an extension of the branch upon which it sits.

The tawny frogmouth resembles the potoo in being disguised as a stump of wood and behaves in a similar way when danger threatens.

of wood and bark left behind by feeding elephants. Such species are seldom seen unless first disturbed, and it is usually possible to observe them on the ground only by marking nests carefully when they are discovered and returning to them later.

Though there are fewer cryptic birds in forest and woodland than in open country, superb examples of true disguise are provided by potoos in South and Central America and frogmouths in South-east Asia and Australia. Both types of birds, which are nocturnal and related to nightjars, roost in trees by day, and resemble the broken stump of a branch. They lay and incubate their eggs in the same sort of position as that in which they roost. Frogmouths, for example, lay their single egg on a downy pad on the upper surface of a branch, incubating it with their body orientated along the branch rather than across it. When disturbed by human observers they are usually found in a wonderfully cryptic posture with their head and body held stiffly upright and stick-like, and their eyes closed to a narrow slit. It is said that frogmouths adopt this posture only when danger threatens, and that they have a special alarm call which signals danger to other members of the species. A similar branch-like posture is adopted by roosting woodland owls.

The nests of most birds are hidden behind dense foliage or in holes, rather than disguised, but there are some nests which are exposed and can be said to be adventitiously disguised. For example, many African flycatchers, notably puffbacks, wattle-eyes and the blue flycatchers, build delightful little nests covered in lichens and moss, and placed in the fork of a tree so that they appear to be a continuation of the branches supporting them. The long-tailed tit sometimes builds a similar nest. The pendulous nests of a number of African forest sunbirds are also marvellous examples, for they look just like the bundles of dead leaves and twigs which frequently hang from lianas and thorny climbers. Another good example is provided by the floating weed nests of grebes.

So far no mention has been made of aggressive camouflage or disguise in birds. In fact, examples are neither particularly numerous nor striking, probably because bird predators rely mainly on surprise, speed and agility to catch their prey. However, mention must be made of the white colouration of many gulls, terns and other sea-birds that fish for their prey, for there is evidence that white, seen against the sky, is the least conspicuous colour to fish below the water surface.

The snowy owl provides one of the few good examples of aggressive camouflage among birds. Its white plumage is conspicuous only when the last snow patches melt during the short northern summer, a time when young animals, inexperienced and easy to catch, provide abundant prey.

The exquisite lichen-covered nest of the chin-spot puffback flycatcher is well disguised, even though it is in full view. At the least sign of a predator, the parent flycatchers slip away from their nest, relying on its inconspicuousness to protect their eggs and young.

WEAVING/ARDEA

Warning colour

A great variety of animals protect themselves from predators by making themselves extremely unpleasant in some way. Box-fishes, for example, secrete virulent poisons; stink-bugs have a nauseous taste and smell; many caterpillars are covered with poisonous spines or irritant hairs; and bees and wasps inflict a painful and poisonous sting. Noxious animals such as these generally ensure that predators do not mistake them for palatable cryptic species by being as different and therefore as conspicuous as possible. They make themselves conspicuous by having brilliant warning colours which are intended to communicate the message 'Danger! Keep away!' Their colours are mostly bright and saturated and arranged in bold contrasting patterns. Combinations of black, red, orange, yellow and white are particularly common and are seen to advantage in the colouration of skunks, arrow-poison frogs, ladybird beetles and wasps.

The vast majority of the protective devices associated with warning colouration utilize chemical substances which are either poisonous or have a foul taste or smell. Many amphibians, fish and insects secrete a virulent poison from glands in their skin or some other part of their body, thus securing rapid release from any predator which is un-

fortunate enough to bite them or mouth them. There are also fishes and caterpillars whose bodies are covered in spines which ooze poison when they pierce another animal. Such species intergrade with others whose spines, though lacking in venom, are sharp and capable of inflicting painful wounds. Sometimes the spines are loose or brittle, so they snap off and remain embedded in the attacker's flesh, where they cause festering sores.

Some poisonous animals which employ their poisons against their prey as well as predators have

Previous page: This black and yellow box-fish has a typical combination of warning colours, here associated with the virulent poison that it secretes when attacked. Its bony armour makes it difficult to swallow or crush and helps it to survive the initial attacks of inexperienced predators.

Below, right: The bright pink nymph of the stink-bug, *Pycanum rubeus*, from Borneo, secretes a noxious fluid with an unpleasant warning smell. The adult bug has a similar secretion and smell, but is green and much less conspicuous.

The bright colours of this blister beetle warn that it secretes an oily, caustic liquid containing cantharidin, which raises blisters on the skin of any animal that molests it. Only inexperienced predators attack blister beetles; they associate the experience with the beetles' conspicuous colours and later avoid them.

evolved ways of injecting poisons into other animals actively rather than passively. Snakes, spiders and assassin bugs, for example, inject poisons through specialized fangs or mouthparts, while bees, wasps, some species of ant, scorpions and jellyfishes have specialized stinging organs.

Appropriate behaviour is just as important to animals with warning colouration as it is to cryptic animals, though the two types of behaviour have exactly the opposite aims — to increase conspicuousness on the one hand, and to decrease it on the other. Animals with warning colours contrive to make themselves more conspicuous than they already are by congregating in large groups, resting in exposed positions, and being sluggish in their movements. In fact, the vast majority appear more or less indifferent to capture, making no attempt to escape or conceal themselves if they are attacked by a predator, though they often have special warning displays to deter them. Many such animals advertise themselves with non-visual as well as visual signals: bees and wasps make themselves conspicuous by buzzing loudly, as well as by having striking black and yellow bands on their bodies; similarly, the poisonous secretions of many animals have a strong, characteristic smell which

can act as a warning signal over a considerable distance. In fact, the secretion of some animals lacks any real poisonous property, but has a smell which is so repulsive that predators keep away; thus, it provides the actual protection as well as a warning signal. The nauseous secretion from the anal glands of skunks is a good example.

Warning colouration is obviously most effective during the daytime against diurnal predators and for this reason warningly coloured animals are usually active by day, often in contrast to close relatives that are cryptic and edible. There are, however, a few animals with black and white warning patterns that are effective at night, particularly in combination with other warning signals. Porcupines, for example, rattle their black and white quills as well as displaying them visually, and they emit a strong smell. Particularly fascinating are a number of distasteful moths that have warning colours which are effective against birds by day, and warning sounds which are effective against bats by night.

Another characteristic of animals with warning colours is that they tend to be difficult to kill, mainly because of their exceptionally tough skins and tissues which enable them to survive injuries

that would be fatal to comparable cryptic species. Thus butterfly collectors find Danaid butterflies, such as the monarch, difficult to kill by the usual method of pinching the thorax. Unless they are completely crushed, their elastic tissues simply spring back into shape when they are released, and, though temporarily stunned, the butterflies soon recover.

At first sight it is surprising that animals with warning colours should be so difficult to kill, for their poisonous stings, spines, secretions and other devices would seem to be adequate to deter predators from ever attacking them. However, predators do not avoid such animals instinctively; they must learn by experience which animals are palatable and which are best left alone. They avoid animals with warning colours only after they have attacked them, suffered an unpleasant experience, and learned to associate the experience with the bright colours of the animal concerned. Thus it seems that animals with warning colours are specially tough so they can survive attacks by inexperienced predators. A predator is usually quick to discover its mistake, but even a single bite or peck could be fatal to animals with less tenacity of life. The fact that predators have to learn to recognize warning

colouration emphasizes the importance of warning patterns being as striking and memorable as possible.

That predators do not instinctively avoid animals with warning colours has been proved many times in a great variety of species, including mammals, birds, lizards, frogs, toads and fishes. Thus Cott showed that inexperienced toads have no aversion to catching honey bees, but quickly learn to avoid them after being stung. He experimented by placing hungry toads, one at a time, on the alighting platform of a bee-hive. Of 34 toads tested, all except

Like other assassin bugs, *Eulyes amaena*, from Borneo, overcomes its prey by piercing it with its tube-like mouthparts and injecting saliva containing a nerve poison. As a defence against predators, it can spit its saliva accurately over a distance of several inches. If its saliva gets into the eyes, it can cause blindness.

Above, left: This strikingly coloured Turbellarian worm from Malaysia is easily recognized and avoided by predators that have already experienced its nauseous taste.

Below left: By congregating together, the brightly coloured, spiny caterpillars of the Malay lacewing butterfly make themselves even more conspicuous than usual. Black, red and white is one of the most common combinations of warning colours.

Caterpillars of a Malaysian Lasiocampid moth mass together in an exposed position to make themselves conspicuous.

three immediately caught bees as they emerged from the hive, and most showed considerable signs of distress as soon as they were stung. Several toads refused to catch any further bees at all, but others persevered for a short while with waning enthusiasm. No toad needed more than seven 'lessons' and ten needed only one. Furthermore, the 'lesson' was evidently well remembered, for 18 toads were tested again after two weeks, and though they had eaten nothing during this time nine made no attempt to catch even a single bee. The other nine caught only a few bees before refusing to be tempted further. All the toads readily accepted mealworms and other food items, proving that they really were hungry.

STINK-GLANDS AND SPINES IN MAMMALS AND BIRDS

In contrast to other animals, most of the mammals that are warningly coloured are nocturnal, and therefore have black and white patterns that are an effective signal on all but the blackest nights. Black and white warning patterns are most strongly developed in Musteline carnivores, such as skunks, polecats and badgers, in which they are

associated with powerful anal stink-glands, and in porcupines, in which they are associated with a formidable armament of spines. Of course, the conspicuous colouring of these animals prohibits them from catching their food by stealth; most of the Mustelines have a very mixed diet of small easily caught animals and some vegetable matter, while porcupines are entirely herbivorous.

The most notorious of the Mustelines for their unbearable stink are undoubtedly the skunks of North and South America, whose anal gland secretions can be aimed in a fine spray for distances of up to ten feet. Their stink is unbelievably foul and was described by W. H. Hudson in *The Birds of La Plata* as an effluvium 'after which crushed garlic is lavender, which tortures the olfactory nerves, and appears to pervade the whole system like a pestilent ether, nauseating one until sea-sickness seems almost a pleasant sensation in comparison'. Skunks seem well aware of the effectiveness of their armament for they are bold and show little sign of fear of other animals. If anything does disturb them they display their plume-like tail on high like a flag, while the spotted skunk has a remarkable warning display in which it highlights the pattern on its back by doing a handstand.

There are many other Mustelines with black and white or brown and white patterns that have a smell equally or almost as bad as that of skunks, notable examples being the African polecat or zorilla, the ratels of Africa, Arabia and India, the teledu of Indo-Malaysia, and the grison of South America. Like skunks, all these animals are fearless creatures which stand their ground if attacked by predators. Indeed, most have a reputation for ferocity which may be as great a deterrent as their revolting smell. The grison, for example, was described by Hudson as 'malignant and bloodthirsty beyond anything in nature', while ratels are said to be undeterred by any adversary even the size of a buffalo. These Mustelines also tend to have a thick rubbery skin, so loose in ratels and badgers that it is almost impossible for other animals to get a good enough grip on them to avoid getting badly bitten in return by their enormously powerful jaws. In fact, most of them are so tough that they are well able to survive the

In common with other skunks, the hog-nosed species, such as this *Conepatus leuconotus*, have a foul smelling anal secretion which they can spray accurately at any adversary. Though nocturnal, skunks have striking black and white warning patterns that are readily recognizable even at night. They also have warning displays in which they stamp their feet threateningly and flaunt their tails.

attack of even the most powerful predator long enough for it to realize its mistake.

Porcupines have a revolting smell as well; in the case of the North and South American tree porcupines it is described as resembling that of highly concentrated human perspiration. However, their smell may be little more than a warning signal, for their chief armament is a battery of long, sharply pointed quills, concentrated on their back and tail. The quills are usually banded with black and white, and are quite conspicuous at night, particularly when fully erected in warning display. This display is noisy as well as visual, for porcupines squeal and grunt, and the African and Asian species have specially modified quills which act as rattles. If attacked, porcupines present their bristling back and tail to their adversary, and try to drive their quills into its body. Frequently the quills are embedded in the unfortunate victim, for they are easily detached and are barbed in the American species. Predators as large and formidable as lions, leopards and pumas are usually repulsed with ease, though they sometimes persevere and come to grief, and may be completely incapacitated by sores caused by stumps of quills embedded in their face, paws and body.

Warning colouration is almost unknown among birds, though there is some evidence that a few black species, notably wood-hoopoes and drongos, are distasteful to at least some mammalian predators, such as cats and mongooses. Having handled wood-hoopoes, we know that they have a peculiarly obnoxious musty smell which is probably caused by substances secreted from the preen gland, but we have no evidence that it is either effective or ineffective in repelling predators. However, there is no doubt that the overwhelming majority of conspicuous or brilliant colours in birds are social signals, and not warning signals associated with noxious properties.

POISONOUS GLANDS IN AMPHIBIANS

Many frogs, toads and salamanders have glands in their skin which secrete deadly poisons. Associated with these poisons there is usually a bitter or burning taste, or an obnoxious smell, so that any predator foolhardy enough to attack such animals

Like skunks, African polecats or zorillas are protected by a nauseous anal secretion and have a conspicuous black and white warning pattern.

quickly regrets it. Even the merest trace of these poisons is usually sufficient to cause violent vomiting, or death in extreme cases.

Some of the most poisonous species of all are the South American arrow-poison frogs in the genus *Dendrobates* whose poisons, which act on the central nervous system, are so deadly that they are used by Columbian Indians for poisoning their arrow-heads. Species of *Dendrobates* are brilliantly coloured. *D. tinctorius*, for example, has variable patterns of red, yellow or white on backgrounds of black or electric blue. One of the earliest accounts of a poisonous frog was written by the naturalist Thomas Belt in *The Naturalist in Nicaragua*, and admirably described its habits which are typical of those of animals with warning colouration:
'In the woods around Santo Domingo there are many frogs. Some are green or brown, and imitate green or dead leaves, and live amongst foliage. Others are dirty earth-coloured, and hide in holes

or under logs. All these come out only at night to feed, and they are all preyed upon by snakes and birds. In contrast with these obscurely coloured species, another little frog hops about in the daytime dressed in a bright livery of red and blue. He cannot be mistaken for any other, and his flaming vest and blue stockings show that he does not court concealment. He is very abundant in the damp woods, and I was convinced he was uneatable as soon as I made his aquaintance and saw the happy sense of security with which he hopped about. I took a few specimens home with me, and tried my ducks and fowls with them; but none would touch them. At last, by throwing down pieces of meat, for which there was a great competition among them, I managed to entice a young duck into snatching up one of the little frogs. Instead of swallowing it, however, it instantly threw it out of its mouth, and went about jerking its head as if trying to throw off some unpleasant taste.'

When threatened, the East African crested porcupine erects the formidable armament of black and white quills on its back, rattles the specially modified quills in its tail, squeals and stamps its feet. If its warning display is an inadequate deterrent, the porcupine lunges backwards and attempts to ram its quills into its attacker's face.

ROSS

The conspicuous red and black Peruvian frog, *Phylobates bicolor*, is protected by its highly toxic skin secretions. This species carries its tadpoles on its back until they can fend for themselves, a habit made possible by the high humidity of the equatorial rain forest in which it lives.

Right: The distasteful African tree-frog, *Hyperolius marmoratus*, has very variable, but always conspicuous, colours and markings. Unlike cryptic tree-frogs, it is to some extent diurnal and sits prominently on leaves or other vegetation. If alarmed, it displays the red insides of its legs but makes little attempt to escape.

A highly poisonous South American toad, *Atelopus stelzneri*, was described in rather similar terms by Darwin in his book *A Naturalist's Voyage Round the World*:

'If we imagine, first, that it had been steeped in the blackest ink, and then, when dry, allowed to crawl over a board, freshly painted with the brightest vermilion, so as to colour the soles of its feet and parts of its stomach, a good idea of its appearance will be gained. If it had been an unnamed species, surely it ought to have been called *Diabolicus*, for it is a fit toad to preach in the ear of Eve. Instead of being nocturnal in its habits, as other toads are, and living in damp obscure recesses, it crawls during the heat of the day . . .'

Brilliantly coloured poisonous frogs and toads with similar adaptations are found in other parts of the world. The little African tree-frog *Hyperolius marmoratus*, for example, is brightly coloured, diurnal and distasteful to predators. An even better African example is the grey or black *Phrynomantis bifasciata* which is conspicuously marked with vermilion or pink stripes and spots. When alarmed by a predator, it exudes a sticky, milky, poisonous secretion from its skin, which is powerful enough to cause considerable irritation to careless human hands and can have a much more drastic effect on a predator's mouth. Like the South American examples it is a very sluggish species which freely exposes itself by day.

It is curious at first sight that there are a considerable number of poisonous frogs and toads that are not particularly conspicuous in their behaviour and colouring, and even some that are extremely cryptic. Good examples are the huge South American toads, *Ceratophrys cornuta* and *Bufo marinus*. *Bufo marinus* has poison so virulent that it is said to be responsible for the deaths of many dogs that are unfortunate enough to molest it. They are cryptic because they ambush their prey and warning

colouration is incompatible with such a method of hunting. Species such as these have to rely on special displays to advertise their poisonous nature: *Ceratophrys cornuta* and *Bufo marinus* puff themselves up to an enormous size and utter warning cries, while other species have concealed patches of colour which can be suddenly revealed. The European fire-bellied toad adopts a peculiar posture with its head held high, displaying its red, white and black belly; sometimes it even rolls on its back as does also the black and white-bellied American toad *Bufo americanus*. A South-east Asian species, *Callula pulchra*, inflates itself with air, and in so doing exposes broad yellow stripes which are normally concealed by folds of loose skin on its back.

Among other amphibians, some of the salamanders are brilliantly coloured and notorious for the potency of their milky, skin secretions. Thus the fire salamander of Europe, which is coal-black and chrome yellow, has a poison capable of killing any small predator. However, while brilliant colours in frogs and toads are a sure sign that they are poisonous, this is not necessarily the case in salamanders. The group of European species known as newts, in contrast to all frogs and toads, have elaborate visual courtship displays, and their pretty colouring is probably for purely social purposes.

POISONOUS SECRETIONS AND SPINES IN MARINE FISHES

As in the case of birds, there is little doubt that the majority of conspicuous and brilliant colours in fishes are social signals, not warning signals associated with poisonous or other unpleasant properties. There are, however, notable exceptions, particularly among marine reef fishes. The brilliantly coloured box-fishes secrete a virulent poison when attacked by a predator, and some puffer-fishes are known to have poisonous flesh. Nevertheless, puffer-fishes are esteemed as a delicacy by the Japanese, though only after special preparation in which the blood-vessels are removed. Unlike other brilliantly coloured reef fishes, box-fishes and puffer-fishes exhibit the fearless sluggish behaviour typical of animals with warning colours, and are well adapted to survive the initial attack of an inexperienced predatory fish. Puffer-fishes can inflate themselves with air or water, and thus enlarged are difficult to swallow, especially in the

BISSEROT/COLEMAN

The chrome yellow and black warning colours of the European fire salamander are associated with an extremely poisonous milky skin secretion.

Right: Well protected by a venomous dorsal spine, species of catfish in the genus *Plotosus* swim in dense shoals and make no attempt to avoid predators.

Puffer-fishes are quickly discovered to be poisonous by any inexperienced predator that attacks them; they generally manage to survive such attacks by inflating themselves with air or water, so that they become difficult to swallow. Like many other animals with warning colours, they are sluggish in their movements and feed on sedentary or slow moving animals.

POWER/COLEMAN

case of the species that are also spiny. Box-fishes are an equally difficult mouthful, enclosed as they are in a hard casing of bony plates, through which only their tail, fins and jaws project and are free to move.

Some marine fishes illustrate the way in which animals with warning colours often congregate to enhance the effectiveness of their warning displays. The case of the catfish, *Plotosus anguillaris*, from the Phillipines offers a particularly superb example. It was originally described by T. Mortensen:
'On the flats of the coral reef at Little St. Cruz Island in the Strait of Basilan, near Zamboanga, my attention was attracted by a very conspicuous black thing moving about in the shallow water. On coming close to it I saw it was a mass of small fishes, black, with two longitudinal white stripes on the sides of the back. They swam very close together,

making thus a large ball; by the constant movement of the small fishes among one another the ball seemingly rolled along over the corals, making an exceedingly conspicuous object. It was quite easy to catch nearly all of them with a single stroke of a handnet. In order to preserve some specimens of them I started to take them with the hand from the net. The first one I touched stuck to my fingers, producing a most intense pain, and on trying to get it off, I had it hanging in my other fingers. It was exceedingly painful, the pain lasting quite a while after I had succeeded in getting it off. After this experience I avoided, of course, most carefully to touch any specimen of this fish, and when I had succeeded in getting some of them preserved I kept carefully away from these black, rolling balls.'

When molested, the African grasshopper, *Dictyophorus*, displays its red and black hindwings and, with a hissing noise that can be heard several yards away, expels a strong smelling, noxious froth from its thoracic spiracles.

Above, right: Like species of *Dictyophorus* and *Phymateus*, the brightly coloured African grasshopper, *Zonoceros elegans*, secretes a noxious froth from its thoracic spiracles, rests in exposed positions, moves sluggishly and makes little attempt to escape from predators. This Ugandan form is winged, but some adult *Zonoceros elegans* are wingless.

Right: The discharge of repellent froth by the African grasshopper, *Phymateus viridipes*, is accompanied by a spectacular warning display.

Some scorpion-fishes, with their poisonous spines modified from dorsal fin-rays, also have warning colours: several species, including the well known lion-fish, are strikingly patterned with red and black or pink and black and ornamented with enormous fan-like fins. They are sometimes described as cryptic when seen against the background of a coral reef, but they are really most conspicuous, behaving in a distinctive manner which is the very opposite of cryptic. Like box-fishes and puffer-fishes, lion-fishes sail securely through the water with complete disregard for predators, though they intensify their colour and display their fins aggressively if approached too closely. It is true that their eyes are disruptively camouflaged, but this may well be protection against sabre-toothed blennies, which are liable to attack their eyes.

However, some members of the scorpion-fish family genuinely are cryptic. The notorious stone-fish is a good example and is quite different from species like the lion-fish in its appearance and behaviour. It is not entirely clear why the stone-fish combines poisonous spines with cryptic, rather than warning, colouration, but it is probably because it ambushes its prey and cannot afford to be conspicuous. This is probably true of sting-rays also, for they too possess a protective poisonous spine in combination with cryptic colouring.

NOXIOUS SECRETIONS, SPINES AND HAIRS IN INSECTS

There are so many insects which protect themselves by secreting poisonous or obnoxious fluids from specialized glands, that it is impossible here to mention more than a few of the most striking examples. Among the best are a great variety of shorthorn grasshoppers, particularly in the genera *Dicty-*

ophorus, *Zonoceros* and *Phymateus*, which discharge an offensive foam from their thoracic spiracles when they are bothered by a predator. All the species concerned are brilliantly coloured, diurnal and exceedingly sluggish by comparison with cryptic grasshoppers. Many species make themselves more conspicuous by congregating in dense masses, particularly in their nymphal stages; others fan out their vividly coloured hindwings.

W. A. Lambourn has described a kitten's first encounter with a specimen of *Zonoceros elegans*:
'My kitten seized one in its mouth, but dropped it instantly, and commenced to salivate. Strings of thick mucus an inch long appeared at the corner of its mouth with much frothy saliva, and the animal retched but did not actually vomit. It kept on rubbing its mouth against a tree in an endeavour to get rid of the cause, and was not its usual self for two or three hours. The Acridian had hopped off, apparently little the worse for its experience.'

The monarch butterfly and other species in the Danainae, Acraeinae and Heliconiinae discharge repellent fluids from their wing-veins, leg-joints and antennae; tiger moths, burnets and many other species in the Arctiidae, Zygaenidae and Syntomidae discharge them from cervical glands; many caterpillars from their mouths; and ladybird beetles from pores around their leg joints. All these insects exhibit conspicuous colours and ostentatious behaviour like the shorthorn grasshoppers mentioned above. The North American monarch butterfly and many ladybird beetles are particularly interesting in that they mass together to hibernate: thus many thousands of monarchs migrate every autumn to Pacific Grove near San Francisco, where they congregate on the same small group of trees. The resulting concentration of visual and scent signals must have a tremendous warning impact on the predators in the area. The smell of the monarch is of great interest for it has been suggested by Miriam Rothschild that it might have an hallucinogenic effect on predators, and invoke vivid recall of any previous occasion on which they have come across it. In other words, it might speed up the learning process by which a predator comes to associate the warning signals with previous bad experiences with monarchs.

The black, yellow and white caterpillars of the common European magpie moth have an exceedingly distasteful secretion in their skin. A. G. Butler discovered this in the middle of the last century in experiments using lizards, frogs and spiders as predators. In the case of frogs Butler found that: 'When they first became aware of the introduction of the caterpillars, they seemed greatly excited, sprang forwards, and licked them eagerly into their mouths; no sooner, however, had they done so, than they seemed to become aware of the mistake that they had made, and sat with gaping mouths, rolling their tongues about, until they had got quit of the nauseous morsels, which seemed perfectly

The garden tiger moth and other Arctiid moths manufacture their own poisons and accumulate additional ones when their caterpillars feed on poisonous plants. To advertise their noxious qualities, tiger moths display their colourful wings and abdomen and bubble a poisonous secretion with a warning scent from their prothoracic glands.

Every year thousands of North American monarch butterflies overwinter on traditional 'butterfly trees' at Pacific Grove in California. By congregating in this way, the butterflies come into contact with fewer predators and their warning colours and scents reinforce each other.

Like the North American monarch butterfly and a number of other insects with warning colours, ladybird beetles mass together to sleep through adverse cold or hot, dry conditions.

When attacked, many of the hairy caterpillars of Lasiocampid moths expose conspicuous velvety patches of highly irritant hairs and arch backwards attempting to drive the hairs into their attacker.

Above, left: The spines of the lurid pink slug caterpillar of the Limacodid moth, *Thosea vetusta*, from Borneo, can cause excruciating pain if touched.

Right: The caterpillars of many Lasiocampid moths mass together, but only become really conspicuous when they are alarmed by a predator. When this happens, they reveal a patch of coloured irritant hairs and lash about. The green droplet discharged from the mouth of this East African species is probably harmless, but it has a warning effect because it resembles the poisonous droplets discharged by many other insects.

uninjured, and walked off as briskly as ever. After this, it was useless to attempt to persuade the frogs to touch one of these caterpillars.'

And in the case of spiders:

'I repeatedly put them into the webs both of the geometrical and hunting spiders (*Epeira diadema* and *Lycosa* sp.); but in the former case, they were cut out, and allowed to drop; in the latter, after disappearing in the jaws of their captor down his dark silken tunnel, they invariably reappeared either from below, or else taking long strides up the tunnel again.'

These are interesting instances of the ability of noxious animals with warning colours to survive the attack of predators.

Many caterpillars are densely covered in brightly coloured hairs or spines. Often these afford purely mechanical protection, but many species have hollow hairs or spines which connect with poison glands at their base. Particularly virulent species include the caterpillars of South American flannel moths and silkmoths in the genus *Automeris*. In Brazil the flannel moth caterpillars are known as *bizos de fuero* ('fire beasts') and sometimes cause paralysis for several weeks in particularly sensitive people. Equally bad are the widely distributed slug caterpillars of Limacodid moths; we can vouch from personal experience for the excruciating pain caused by contact with the spines of some of the South-east Asian species. Also unpleasant are the hairy caterpillars of Lasiocampid moths, many of which have strikingly coloured patches of minute urticating hairs concealed under folds of skin on their back. If attacked, some species arch backwards and attempt to drive the fragile, sharply

pointed hairs into their assailants, much as porcupines try to drive their quills into larger predators. Other species, such as the processionary caterpillars, have hairs which are detached and dispersed by specially modified long, branched hairs. When thousands of caterpillars display together, as often happens, the air is filled with the tiny hairs, which cause intense itching, and sometimes dangerous inflammation in the eyes, throat and nose of their attackers. Most of these species make use of their urticating hairs even after pupating, weaving them into their cocoons, where they continue to provide effective protection against predators.

Just as there are insects which parallel the protective adaptations of porcupines, there are others that parallel skunks by being able to spray ob-

The East African cuckoo wasp, *Stilbum cyanurum*, has brilliant metallic green warning colours. Like all cuckoo wasps, it is a solitary species that parasitizes other wasps and bees; its exceptionally tough outer 'skin' protects it from the sting of its host.

noxious secretions over a distance of several inches. The classic example is the bombardier beetle which ejects a spray of caustic fluid from the tip of its abdomen. It actually aims the spray at attacking predators, and is capable of emitting up to 20 successive discharges, each of which is brought about by mixing two chemicals that are normally stored separately. An explosive reaction results from this mixing and the oxygen that is liberated provides the pressure for the discharge of the caustic spray.

POISONOUS STINGS IN HYMENOPTERA

A number of animals use poisons to overcome their prey as well as to defend themselves and have evolved ways of actively injecting the poisons into their prey and predators. Bees, wasps and hornets, with their specialized stings, are notable examples. The stings are modified parts of the female reproductive system and are consequently lacking in males. Many of these species are social insects which live in large conspicuous aggregations and all are conspicuous in their behaviour and colouring. Some are completely black, others are arrayed in beautiful metallic reds and greens, while others have black and yellow, or brown and yellow, banded

bodies and are among the classic examples of warning colouration. The sting of the social bees is used only for defensive purposes, but other species, particularly the solitary wasps, use their sting to paralyse the spiders and caterpillars which they store in their nests to feed their larvae. Some of the biggest of the spider-hunting Pompilid wasps are particularly awesome insects; they reach over three inches in length and are capable of overcoming even the largest of the bird-eating spiders of tropical forests. However, not all bees and wasps have effective stings. The annoying small, black, sweat bees of the tropics, for example, have no sting to speak of, but do have unpleasant secretions which deter most predators.

The Hymenoptera also include ants and many of these, notably in the primitive Ponerine group, have stings similar to those of bees and wasps. Many ants, however, have no sting and rely instead on other deterrents that are no less effective: the red tree-ants of Malaysian forests bite savagely and secrete formic acid over the resulting wounds, while other species defend themselves by squirting a spray of formic acid at their adversaries. The common European wood ant is remarkable for its ability to store formic acid at concentrations of up to 70 per cent and was utilized by man as a source of this acid from Roman times until fairly recently.

POISONOUS STINGS IN
MARINE COELENTERATES

Animals with specialized stinging organs are very abundant in the sea in the form of jellyfishes, sea-anemones and other Coelenterates. Most are brightly coloured and many emphasize their unpleasant nature by phosphorescing when the water around them is agitated by a large and possibly dangerous animal. Their stinging organs, which are used for catching prey as well as for defence, are specialized cells called nematocysts, distributed in enormous numbers along their tentacles. The nematocysts are of several different types, but all contain a thin thread which is ejected when the cell's sensitive trigger-hair is touched by another animal. The stinging type of nematocyst has a sharply pointed, barbed thread which pierces the animal at which it is ejected and discharges a poison.

Jellyfishes are often a considerable hazard around coasts all over the world. Many species trail long tentacles for several yards behind them, and it can be agonizing to brush against them and

receive the discharge of hundreds or even thousands of nematocysts. The colonial Siphonophoran jelly-fishes, such as the notorious but beautiful Portu-guese man-o'-war, use particularly virulent poisons which occasionally kill bathers. The stinging cells of the Portuguese man-o'-war have quite a different effect from those of typical Scyphozoan jellyfishes. Even tiny specimens of the former produce glandu-lar swelling and a peculiar sharp pain which con-tinues unabated for several hours, before stopping, quite suddenly and completely, within minutes. A small jellyfish merely inflicts a fairly short-lived pain like a nettle sting and there is no accompany-ing glandular reaction.

WARNING COLOURATION AND
ADVENTITIOUS PROTECTION

Many plants contain poisons which protect them from being eaten by animals. In spite of this, many of them are exploited by insects which have evolved ways of dealing with the poisons. Some insects, which tend to be cryptically coloured, excrete the poisons with exceptional efficiency. Others, which invariably have warning colours, store the poisons for their own protection, providing a contrast to

Left: The batteries of stinging cells on the tentacles of the violet jellyfish and other Coelenterates have a deterrent effect on preda-tors as well as enabling them to catch and kill their prey.

Like the North American monarch butterfly, the African monarch stores heart poisons from the milkweeds on which it feeds as a caterpillar; its degree of toxicity depends on that of the milkweed species concerned.

the warningly coloured animals that manufacture their own poisons from non-toxic substances in their diet. In so far as they acquire their protection from the environment, these animals have a form of adventitious protection which parallels that of some cryptic animals.

Insects capable of storing plant poisons in their bodies are found in at least six orders, including grasshoppers, plant bugs, butterflies and moths, lacewings, beetles and flies. The most studied examples are probably the butterflies in the Danainae, a pantropical group which includes the familiar monarchs. Monarch caterpillars feed exclusively on one family of plants, the aslepias or milkweeds, which contain heart-poisons chemically similar to digitalis. It is now known that different species of milkweeds vary in their toxicity from being completely innocuous to very toxic, and that the monarch caterpillars which feed on these plants and the resulting adult butterflies show a corresponding gradation in their degree of distastefulness. This discovery satisfactorily accounts for earlier conflicting claims concerning the toxicity of the North American monarch. Some experimenters claimed that they were palatable to predators, others that they were extremely distasteful; in fact, both claims were correct.

Another group of species which vary in toxicity are the tiger moths in the family Arctiidae. Tiger moths manufacture poisons, but some are also protected by additional toxins obtained from plants. The caterpillars of the European garden tiger moth, for example, feed on a great variety of plants, some poisonous, some not, and are capable of storing completely different poisons. For example, they store digitalis when they feed on foxgloves and toxic alkaloids when they feed on ragwort.

Even more remarkable are certain sea-slugs which protect themselves with the nematocysts or stinging cells of the sea-anemones upon which they feed. The sea-slugs can prevent the nematocysts from discharging, presumably by means of a chemical slime similar to that used by the sea-anemones themselves and by the clown-fishes and other animals that associate with sea-anemones without being harmed. After being ingested, the nematocysts are transferred intact to specialized areas on the backs of the sea-slugs, where they are fully capable of discharging their stinging threads and act as an effective deterrent to predators.

As was mentioned earlier, there are other sea-slugs which incorporate into their skin the pigments and needle-like spicules of the sponges upon

Sea-slugs in the family Doridae browse on sponges and incorporate the pigments and protective, needle-like spicules of the sponges into their own skin. These particular sea-slugs, *Doris*, are feeding on a pink sponge and have become pink.

The brilliant warning colours of *Fascelina coronata* and other sea-slugs in the family Aeolidiidae are associated with the stinging cells in the outgrowths on their backs. These stinging cells are obtained from the coeleterates on which they feed.

which they rest and feed. They provide examples in which both the colouration and the protection are adventitious; their colouration is also both warning and cryptic, for though the sponge has warning colours, the sea-slugs resemble their sponge background so closely that they are effectively camouflaged.

The sponge crabs that carry a piece of living sponge on their back provide another example that could be classified equally well as either adventitious disguise or adventitious warning colouration. It is also an example of a mutually beneficial partnership, for the sponge gains the advantage of

being carried to new feeding grounds. Admittedly, their particular partnership is rather one-sided; the crab is completely dependent upon the sponge, never being found without it, while the sponge is an involuntary partner and can live perfectly well on its own. In other cases, dependence is more mutually advantageous. Some crabs for example, are protected by specific warningly coloured sea-anemones which are carried on their backs. The hermit crab, *Eupagurus prideauxi*, and the anemone, *Adamsia palliata*, for example, are almost always associated with each other and are found apart only when very young. The sea-anemones benefit by being able to feed on scraps of food scattered in the surrounding water by their host.

Just as some marine animals gain adventitious protection by associating with stinging Coel-

The sponge crab, *Dromia*, gains protection from the piece of living sponge that it carries on its back.

enterates, some terrestrial animals gain protection by associating with stinging Hymenoptera. The most notable are the nesting associations between many tropical birds and wasps. The birds concerned are generally species, such as weavers, mannikins and Icterids, which build large conspicuous nests, while the wasps with which they choose to associate are almost always the most aggressive and venomous in the area. Birds also nest in association with some of the more unpleasant tropical tree-ants. However, it is doubtful whether the wasps and ants benefit from these associations and it is not at all clear why they allow the birds to remain so near their own nests.

THE LIMITATIONS OF NOXIOUS PROPERTIES AS A MEANS OF DEFENCE

For an animal with warning colours it is important that its toxic or nauseous properties should be neither so strong as to be lethal, nor so weak as to be ineffectual. If predators are killed at their first encounter, they have no chance to learn by experience and their place will be taken by another inexperienced predator. Hence, even if the animal with warning colours survives the initial encounter, it runs the risk of a succession of further attacks and is unlikely to survive them all. It is preferable for the original predator to survive, for thereafter it will avoid the warningly coloured animal and, if it is territorial, prevent other inexperienced predators of the same species from entering the area. Indeed, predators generally do survive, even after consuming an extremely toxic animal, such as a monarch butterfly, because the poisons concerned are usually combined with an emetic; predators rid themselves of most of the poison by vomiting, and suffer only short-term ill-effects. The only poisonous animals that regularly cause death are snakes, scorpions, and spiders, whose poisons are injected and used primarily to kill their prey rather than as a defence against predators. It is significant that the majority of the most venomous snakes avoid encounters with predators by being cryptic, while the few that do advertise themselves with warning colouration are usually mimics of less venomous species. The significance of warning colouration and mimicry in snakes is discussed in more detail later.

It is equally important that the defences of animals with warning colours should not be so weak as to be ineffectual for, by advertising themselves so conspicuously, they are automatically vulnerable to any predators prepared to eat them. A critical factor here is the availability of alternative prey, for predators are less choosy when food is scarce and will eat animals that are avoided at other times. For this reason a number of mildly distasteful tropical butterflies have seasonal colour forms. The African pansies, for example, have a generation with warning colours in the wet season, when food for predators is abundant, and a cryptic generation in the dry season, when food is relatively scarce.

Animals with warning colours are destined to remain scarce in comparison with cryptic species, for predators would otherwise evolve adaptations enabling them to exploit such a conspicuous and readily available food source. In fact, a few predators do specialize in feeding on certain relatively abundant, nauseous or poisonous animals: cuckoos eat conspicuous hairy caterpillars with relish; bee-eaters have specialized behaviour for de-stinging bees and wasps; honey badgers have hides that are relatively impervious to the stings of bees; anteaters, pangolins and many woodpeckers are physiologically adapted to consume and digest ants that are protected by concentrated formic acid; and some mice are said to destroy the 'bombarding' apparatus of bombardier beetles by rubbing the end of their abdomen on the ground. Particularly interesting are the many hornbills that have evolved specialized techniques for dealing with the poisonous snakes, centipedes and scorpions that make up a large part of their diet. The poisonous animal is held in the very tip of their long bill and repeatedly squeezed along its whole length, as it is moved back and forth. As each end of the animal is reached, be it head or tail, it is given a particularly vicious squeeze. The process is repeated many times, depending on the size and hardness of the animal. The survival value of this behaviour is clear, for it ensures that the dangerous stinging or biting apparatus of the animal, in the head of a snake or centipede or in the tail of a scorpion, is completely crushed. This behaviour can be seen both in wild hornbills and in tame free-flying species. In Sarawak our tame hornbills treated any long flexible object, such as a piece of thin rope, in precisely this way. Both wild and tame hornbills also spend much time 'playing' with twigs and leaves, tossing them in the air, catching them, and passing them backwards and forwards in their bill. Such play can be regarded as practice for a feeding method that demands great skill and dexterity.

The poisonous caterpillars of the cinnabar moth have black and yellow bands similar to those of many wasps. The caterpillars and the wasps are Müllerian mimics, for predators that have learned to avoid either subsequently avoid both.

Left: The African monarch (below) and this species of *Acraea* from Nigeria are Müllerian mimics; in spite of the fact that they belong to different subfamilies, they resemble each other almost exactly in colour, pattern and behaviour, and both are distasteful to predators.

MULLERIAN MIMICRY

The German naturalist, Fritz Müller, who lived and collected butterflies in Brazil, was the first to notice that different species of distasteful butterflies often have extremely similar warning patterns. He realized that this mutual resemblance is an advantage to all the species concerned, for predators have to learn to avoid only one warning pattern instead of several. Hence, the attacks made by predators while still learning to avoid the pattern are shared between all the species, each of which suffers fewer losses on average than it would if the predator had to learn to avoid several distinctive patterns. This mutual resemblance of warningly coloured noxious species has come to be known as Müllerian mimicry and is very widespread, particularly among insects.

Müllerian mimicry is by no means restricted to closely related species, such as different species of butterflies or different species of beetles. Insects as different as wasps and cinnabar moth caterpillars share a similar yellow and black banded pattern and are Müllerian mimics. It has been shown experimentally that bird predators, having learned to avoid wasps, also refuse to touch cinnabar moth caterpillars, even though they have not experienced them before. Similarly, birds that have learned to avoid cinnabar moth caterpillars also avoid wasps. This response on the part of the predators depends on a concept known as stimulus generalization. Having learned to associate unpleasant experiences with a particular pattern, the predator plays safe, generalizes, and subsequently avoids animals with similar as well as identical patterns. The extent to which the predator generalizes probably depends on how nasty its previous experiences have been. It is possible, for example, that animals that have been very badly stung by black and yellow wasps would thereafter be so cautious that they would avoid even black and yellow frogs or black and yellow salamanders. If salamanders and wasps are rather unlikely examples of Müllerian mimicry, there are many examples of less widely differing species that are not. Animals as diverse as grasshoppers, caterpillars, Syntomid moths, assassin bugs, Fulgorid bugs, Pentatomid bugs, blister beetles, millipedes and Turbellerian worms can be seen with similar warning patterns in the same South-east Asian rainforest environment. There is no proof that all of them are Müllerian mimics, though it is likely that they are to at least some predators.

Perhaps the most superb examples of Müllerian mimicry are found among the passion-flower butterflies of South and Central America. Two species in particular, *Heliconius melpomene* and *H. erato*, occur side by side over a vast area and resemble each other so closely that they can only be distinguished visually by counting the little red spots on the underside of the hindwing adjacent to the body; *H. erato* has four red spots, *H. melpomene* only three. The two species are, however, easily distinguished by their strong and characteristic smells, described by John Turner, who has made an intensive study of these butterflies, as resembling witch-hazel in the case of *H. erato* and 'a plate of fried rice that has been allowed to get dusty' in *H. melpomene*. Probably the butterflies themselves rely on their sense of smell to distinguish members of their own species, for it is doubtful whether they are able to recognize their own kind visually. The most remarkable thing about these two species is that they resemble each other throughout their geographical range, in spite of the fact that each is divided into a large number of very different geographically replacing colour forms. Along the coast of South America, from Venezuela to French Guiana both species occur in a form which is

Heliconius melpomene aglaope

Heliconius erato estrella

H. m. nanna

H. e. phyllis

H. m. cythera

H. e. cyrbia

H. m. plesseni

H. e. notabilis

The two Central and South American passion-flower butterflies —*Heliconius melpomene* and *H. erato*—are among the most remarkable of all Mullerian mimics. They resemble each other throughout their range, in spite of the fact that both are divided into a large number of geographical colour forms.

When alarmed by a predator, the crested rat parts the fur on its flanks, revealing a black and white striped pattern superficially like that of the fierce, foul smelling, African zorilla. As a result of this resemblance, which is known as Batesian mimicry, predators with prior experience of the zorilla avoid the crested rat, even though it is harmless.

completely black except for a red patch on the forewing, whereas along the lower Amazon both occur in a form which is black with red stripes at the base of both pairs of wings and yellow spots on the forewing. Until recently, several of these different geographical colour forms were assumed to be separate species, but they are now known to hybridize in areas where the edges of their ranges overlap. Though they are in contact now, they must have evolved in isolation and must indeed have been close to becoming separate species. Incidently, these butterflies exhibit all the characteristics of typical warningly coloured animals: they have a leisurely floating flight which shows off their brilliant warning colouration to advantage; they are distasteful; they have a strong smell that probably acts as an additional warning signal; and they frequently congregate at dusk to sleep.

In many areas there are other distasteful butterflies that resemble these Heliconids, including species from such families as the Danainae, Acraeinae and Ithomiinae. Together these species can be thought of as belonging to a mutually beneficial Müllerian mimicry 'club'.

The smells of the passion-flower butterflies are different from each other because they provide the means by which the butterflies recognize members of their own species. However, the smells of many animals with warning colours are very similar and it is possible that there is a kind of Müllerian mimicry involving odour. Ladybird beetles and tiger moths, for example, have the same quinolene-like scent and most Pentatomid bugs have a noxious 'bitter-almond' smell. A form of Müllerian mimicry involving odour would be particularly useful to animals vulnerable to nocturnal predators.

Before the discovery of Müllerian mimicry, the English naturalist, H. W. Bates, discovered a different type of mimicry which has come to be known in his honour as Batesian mimicry. Like Fritz Müller, who followed after him, he gained inspiration for his discovery from his studies of Brazilian butterflies. He noticed among the abundant warningly coloured passion-flower butterflies small numbers of unrelated Pierids that were almost identical in colour and pattern. He was intrigued by the striking dissimilarity of these Pierid butterflies to others in the area, which had the white and yellow colours that are typical of Pierids all over the world. He realized that these atypical Pierids were 'sheep in wolves' clothing', really quite palatable, but protected by their resemblance to the distasteful passion-flower butterflies. The latter belong to the Müllerian mimicry 'club' which also includes other warningly coloured butterflies in the families Danainae, Acraeinae and Ithomiinae, all of which, so to speak, 'sail under the same colours'. By joining this 'club' under 'false colours', the similarly coloured Pierids derive the benefits of membership though, because

CALLOW/NHPA

The hoverfly (above), wasp beetle (below, left) and hornet clearwing moth (below) are Batesian mimics of wasps and hornets. All are harmless, but tend to be avoided by predators that have been stung by a genuine wasp or hornet. Though most predators play safe in this way, some have remarkable powers of discrimination and regularly pick out and feed on Batesian mimics. North American cliff swallows and purple martins are even capable of distinguishing between stingless drone honey bees and workers.

FLETCHER/NSP

BRADLEY/NSP

they are not distasteful, they contribute nothing to the mutually beneficial relationship that exists between the genuine members. In fact, they are positively detrimental, for any predator that eats one will find it palatable, and be encouraged to try other butterflies with similar patterns. Incidentally, Müllerian mimicry is beneficial to predators as well as to the mimics, for the predators suffer fewer unpleasant experiences in learning to avoid one pattern, than they would if they had to learn to avoid several. By contrast, Batesian mimicry is a great disadvantage to predators, for they are deceived into avoiding palatable animals.

Batesian mimicry is widespread among insects. There are a vast number of noxious species that act as models and an equally large number that are palatable mimics, some of them closely related, some not. Bees, wasps and hornets are mimicked by numerous palatable bugs, moths, beetles and flies; Ichneumon wasps by cockroaches, bugs and flies; ants by spiders; bombardier beetles and tiger beetles by grasshoppers and crickets; and ladybird beetles by cockroaches.

The few examples of Batesian mimicry among vertebrates include the crested rat, which mimics the African zorilla, and various black birds, which mimic the African drongos. In the case of the drongo mimicry, Cott described supporting experiments carried out by C. F. M. Swynnerton, who tested a cat's reaction to several potential mimics, including one that was not completely black:
'Swynnerton offered his cat, which had developed an intense dislike for drongos, a drongo, black flycatcher, male cuckoo-shrike, and tit, in that order, and *belly upwards*. They were all ignored. The birds were then turned over, in the same order. The cat still refused the first three. But when Swynnerton turned over the tit, thus displaying the white dorsal markings, the animal at once came forward and tested it.'
As far as the cat was concerned the mimicry was effective in species that appeared to be completely black, like the drongo.

Being a Batesian mimic of a bumble bee, this harmless East African beefly is avoided by most insectivorous birds.

The nymph of the Malaysian katydid, *Condylodera* (above), is a Batesian mimic of the tiger beetle, *Tricondyla* (below), which has powerful jaws and is able to resist other insect predators. *Con-* *dylodera* is so similar to its model that it was first discovered accidentally in a museum collection of tiger beetles.

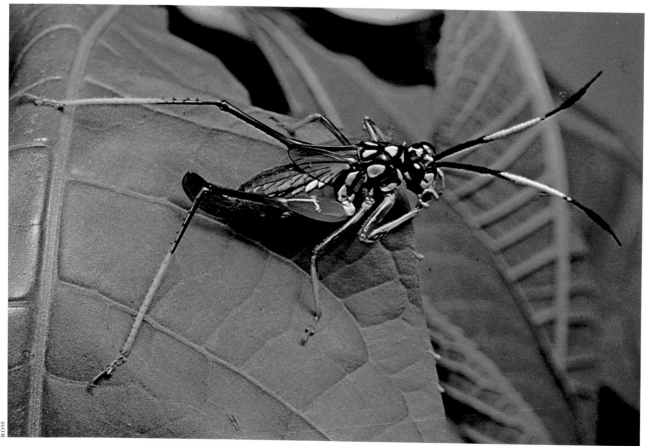

The South American katydid, *Aganacris* (above), is a Batesian mimic of the Ichneumon wasp, *Cryptanura* (below). Like most other Batesian mimics, the katydid resembles its model in its behaviour as well as its shape and colouring; it runs jerkily and constantly moves its antennae in exactly the same way as an Ichneumon wasp.

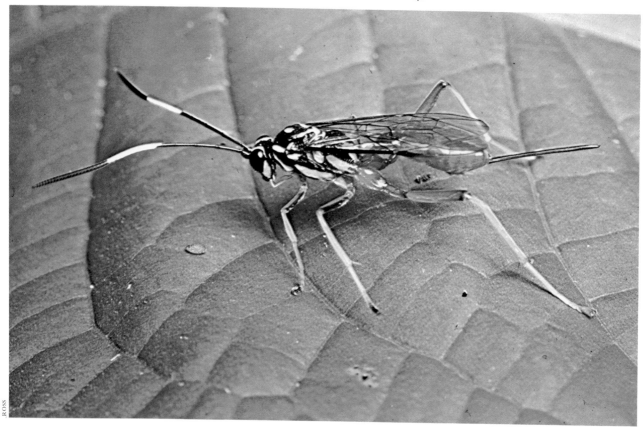

There have been numerous other, more rigorous experimental demonstrations of the effectiveness of Batesian mimicry, notably by L. P. and J. v. Z. Brower. In one such demonstration they used an American bumblebee, its robberfly mimic, dragonflies as control food, and common toads as predators. The results of their experiments are best given in their own words:

'The experiments on the bumblebee/robberfly complex were conducted with six toads. Of these, three were experimental animals and three were control animals. The experimental toads were given ten live bumblebees and ten dragonflies, singly at the rate of four insects a day. At first, each readily seized a bumblebee, but in so doing was severely stung on the tongue and roof of the mouth. The toad reacted by making violent movements with its tongue, by blinking, by listing towards the side of the injury, by puffing up the body, and by ducking the head, which produced a generally flattened appearance. After one or two such experiences, the three experimental toads learned that bumblebees were noxious and they would not strike at the others that were offered. They consistently ate the dragonflies, however, showing that they could distinguish between noxious and edible insects. The crucial part of the experiment was the substitution of robberfly mimics for bumblebees. Would the toads eat the flies, or would they confuse them with the bumblebees and reject them? Of the thirty robberflies that were presented to the three experimental toads, only one robberfly (3 per cent) was eaten.'

At the same time, the Browers showed that the control toads, which had no previous experience of bumblebees, ate robberflies with relish. The experimental toads can only have rejected the robberflies because of their close resemblance to the bumblebees that they had learned to avoid.

The resemblance between many Batesian mimics and their models is extraordinarily exact. Particularly striking in this respect is the close resemblance between distasteful monarch butterflies of several species and palatable butterflies in a number of other families, notably the Papilionidae and Nymphalidae. However, the resemblance between model and mimic need not be exact, for predators generalize in their rejection of Batesian mimics, just as they do in the case of Müllerian mimics. Again, the degree of generalization depends on the previous experiences of the predator, its hunger, and the abundance of alternative prey, but in some circumstances even the slight resemblance of a poor mimic to a very noxious model is enough to protect it from the attacks of predators. For example, the Browers found that birds which had learned to avoid the 'unpalatable red-banded black butterfly *Heliconius erato* subsequently refused to take uniformly black butterflies of about the same size, and also rejected red-banded black butterflies of a completely different shape'. They even suggested 'that some birds under certain circumstances will, after an experience with a butterfly as unpalatable as the monarch, temporarily give up eating all species of butterflies'.

Batesian mimics must, of course, occur in the same general habitat and be active at the same times of day as their models, so that they are exposed to a similar range of predators. In the case of insects which are on the wing for only part of the year, it is also essential that the Batesian mimics should be active in the same season as their models. It would however make sense for the mimics to emerge slightly later than the models, at a time when predators will already have learned to avoid the warning pattern that is common to both.

Batesian mimics resemble their models in many other aspects of their general behaviour. In fact, their behaviour is usually typical of animals with warning colours and quite untypical of the group to which they are most closely related. For example, the leisurely floating flight and unconcerned behaviour of Danaid-mimicking Nymphalid butterflies resembles that of their Danaid models and is completely unlike that of related, but non-mimetic, Nymphalids. Similarly, ant-mimicking spiders run about like ants and not like the jumping spiders to which they are related. They even move their extra front pair of legs in the same way as ants move their antennae. Batesian mimicry sometimes involves the mimicry of sounds as well as visual signals. Hoverflies, beeflies and mantisflies, for example, buzz angrily like the bees and wasps they imitate, particularly if they are molested, and there are palatable nocturnal moths that imitate the ultrasonic warning sounds made by distasteful species to ward off bats.

It is an observed characteristic of Batesian mimics that there are usually far fewer mimics than models, presumably because predators would be encouraged to feed on them if they were encountered too frequently. However, the ratio of mimics to models must depend largely on the exactness of the mimicry and the extent to which predators are deterred by the noxious qualities of the models. If the mimicry is very exact, and the

Below: The significance of ant mimicry is not always clear, but this jumping spider from Borneo is probably a straightforward Batesian mimic. It resembles a forest ant, *Campanotus*, that is seldom eaten by birds or other predators, even though it is abundant and conspicuous in the forest undergrowth. This spider's resemblance to an ant is very precise, for it even has a constriction in its cephalothorax (combined head and thorax) which gives the impression of an ant-like neck, and it moves its front pair of legs in exactly the same way as a real ant moves its long antennae, constantly touching the ground in front of it. The spider gives itself away only when it jumps from a leaf and suspends itself by a silken thread.

This mantisfly from Borneo resembles a mud wasp, *Stenogaster* (left): it is similar in colour and pattern and, when molested, buzzes angrily and makes mock stinging movements with its abdomen. However, since it is parasitic on *Stenogaster*, its resemblance may have an aggressive as well as a protective function.

models are extremely unpleasant, it is theoretically possible for the mimics to form a high proportion of the combined mimic and model population.

Some idea of how large this proportion of mimics can be comes from studies by the Browers on wild populations of the North American monarch butterfly which, it will be remembered, has palatable as well as exceedingly noxious forms, because its caterpillar feeds on milkweed plants of varying toxicity. To all intents and purposes, the noxious form is a model and the palatable form a perfect mimic, a state of affairs described by the Browers as automimicry. In the wild, the Browers found that the very noxious model could be relatively rare, sometimes accounting for only 24 per cent of the total population. This finding agrees well with the theoretical minimum proportion of models needed to confer a worthwhile level of protection to the mimics. The advantage of automimicry to the monarch butterfly is that it enables it to more than double its numbers by feeding on non-poisonous milkweeds in areas in which poisonous milkweeds are rare, whilst still retaining a substantial, though decreased, level of protection from predators. However, there is an upper limit to the size of the population, for the level of predation increases rapidly when the proportion of the palatable form reaches a certain point.

Some of the most remarkable Batesian mimics are butterflies that have evolved different colour forms, each mimicking a different distasteful species. It is possible that this is one way in which Batesian mimics can increase their total population size, and yet still remain rare by comparison with their models. The classic example of this sort of polymorphic mimicry is the widely distributed African swallowtail butterfly, *Papilio dardanus*, described by E. B. Poulton as 'the most wonderful butterfly in the world', though it is now known that there are other swallowtails equally remarkable in the diversity of their colour forms. The subject of mimicry in *P. dardanus* is too intricate for adequate description here, but readers who wish to pursue this subject further should refer to original papers by C. A. Clark and P. M. Sheppard, who have worked extensively on the polymorphism and genetics of this species. In summary, there are four different female forms of *P. dardanus* living together over much of Africa. One resembles the male and is non-mimetic, while each of the other three resembles with remarkable exactitude a completely different noxious Danaid butterfly. One female form mimics the African monarch, while the other two mimic closely related species in the genus *Amauris*. The reason that males of *P. dardanus* are non-mimetic is perhaps because they need to be conspicuous and easily recognized by the females. Females mate only once and there must be no mistake in their choice of partner; if the male resembled another species there would be a danger of confusion.

We have discussed Batesian and Müllerian mimicry as if they were totally different. In fact, like most biological categories, they intergrade; many mimics can be classified as either one or the other, depending on such factors as their degree of noxiousness, the availability of alternative prey and the predators involved. Consider, for example, two species, one of which is highly distasteful, the other only mildly so. To some predators they may be Müllerian mimics, particularly if alternative prey is abundant; to others the mildly distasteful species may be a Batesian mimic of the other; and to yet others both may be palatable, especially if alternative prey is extremely scarce, or the predators concerned are adapted to deal with them.

WARNING COLOURATION AND MIMICRY IN SNAKES

Warning colouration and mimicry in snakes is unusual in several respects. Unlike most other noxious animals, the majority of poisonous snakes try to avoid coming into contact with predators by being nocturnal and relatively unobtrusive in their colouring and behaviour. They seldom advertise their poisonous nature with warning colours. There are several good reasons for this. Firstly, their poisons have been developed primarily for use against prey, rather than predators, and their hunting methods tend to be incompatible with bright warning colouration. In fact, concealing colouration is vital in such thick-set and slow-moving species as the rattlesnakes and vipers, which ambush their prey. Secondly, there are a number of hawks, eagles and other predators that are well adapted to deal with even the most venomous snakes, all of which are perfectly edible once killed. Bright colouration would only serve to make the snakes more noticeable and therefore more vulnerable to predators such as these. Thirdly, the most venomous of snakes are so deadly that they kill at the first encounter any predator that they manage to bite. Predators therefore cannot learn by experience to associate the appearance of snakes with their venomous nature and there can be no straightforward selection for warning colouration. Some poisonous snakes do, however, have warning colouration for reasons that will become apparent.

As animals have no opportunity to learn to avoid

These African butterflies provide superb examples of polymorphism among Batesian mimics. *Danaus chrysippus, Amauris crawshayi* and *Amauris niavius* are three distasteful Danaid models, each of which is mimicked by a different female form of *Papilio dardanus*. *Danaus chrysippus* is also mimicked by the female of *Hypolimnas misippus*, while *Amauris crawshayi* and *Amauris niavius* are each mimicked by a different form of *Hypolimnas dubius*. The males of *Papilio dardanus* and *Hypolimnas misippus* are non-mimetic.

Papilio dardanus tibullus ♂

Hypolimnas misippus ♂

Danaus chrysippus chrysippus

Papilio dardanus tibullus ♀
form *trophonius*

Hypolimnas misippus ♀

Amauris crawshayi

P. d. tibullus ♀
form *cenea*

Hypolimnas dubius wahlbergi
form *mima*

Amauris niavius

P. d. tibullus ♀
form *hippocoonides*

H. d. wahlbergi
form *wahlbergi*

poisonous snakes, there must obviously be selection in favour of individuals that avoid them instinctively and some animals, including man, but excluding those specially adapted to feed on them, do indeed have an instinctive fear of snakes. When snakes come into contact with such animals, it is clearly an advantage for them to accentuate the fact that they are snakes by means of characteristic displays. Many snakes show their sinuous shape to advantage by writhing about and some nocturnal species have conspicuous banded patterns which emphasize their shape even at night. The Southeast Asian coral snake, *Maticora intestinalis*, for example, writhes when it is molested and turns on its back, exposing the conspicuous black and white banded pattern on its belly. A number of related coral snakes and kraits have similar patterns, or

are completely banded above and below. Other snakes have threatening displays. Cobras rear up and spread a patterned hood; boomslangs inflate their neck and thorax; mambas expand their throat pouch; and many vipers display the conspicuously coloured inside of their mouth. However, the most characteristic displays of venomous snakes involve noises. The vast majority hiss loudly when alarmed; the puff adder puffs; some species rasp their scales together as they writhe; rattlesnakes vibrate the specialized rattle on the end of their tail, while the related pit-vipers vibrate their unspecialized tail among dead leaves.

Less venomous or non-venomous snakes are also feared instinctively by many animals, for they are often similar to venomous snakes in appearance and behaviour: they hiss loudly in the same way

When molested, the nocturnal, poisonous, Malaysian coral snake, *Maticora intestinalis*, makes itself easily recognizable as a snake, even at night, by writhing about on its back, displaying its black and white banded underparts. If surprised by day, it tries to deflect attention from its vulnerable head by raising and wriggling the red tip of its tail.

Right: Though normally shy and retiring, cobras have a spectacular warning display in which they rear up, spreading their patterned hood and hissing loudly. Cobras rely on predators avoiding them instinctively, as their venom is so deadly that predators have no opportunity to learn by experience to avoid them. The venom of cobras and other poisonous snakes has evolved primarily for killing prey quickly, not for defence.

and usually have similar threatening displays. A number of Malaysian keelbacks, for example, flatten their necks in imitation of a cobra's hood and many species display the coloured inside of their mouth. In a sense, slightly venomous and non-venomous snakes are like Batesian mimics, for they are protected by their resemblance to venomous species. They differ from Batesian mimics in that the predators concerned avoid the venomous models *instinctively*, not after having *learned* to avoid them.

Though straightforward selection for warning colouration in very venomous snakes is unlikely, it is perfectly possible in mildly venomous forms, provided they do not need to be inconspicuous for another reason. In fact, mildly poisonous species with warning colouration do exist, notably among the New World coral snakes. A particularly interesting result is that predators that learn to avoid these mildly venomous snakes subsequently avoid any others that resemble them in colour, regardless of whether they are harmless or highly dangerous—so here is one way in which very poisonous species can gain protection against predators which do not avoid them instinctively. This form of mimicry, in which highly venomous species mimic moderately venomous species, was first recognized by Wolfgang Wickler, who suggested that it should be known as Mertensian mimicry, after R. Mertens, who made extensive studies of the New World coral snakes in which it is so prominent.

Mimicry in the New World coral snakes is complicated and controversial, some authorities even denying that it exists at all. Over seventy species of American snakes have variations of the 'coral' pattern of black, red, yellow and white bands. The true coral snakes are highly poisonous and belong to the Elapidae, a family that also includes such deadly species as cobras, kraits and mambas. The false coral snakes belong to the ubiquitous family Colubridae, which includes species ranging from harmless to moderately poisonous. Species from the two families are sometimes so similar that they are difficult to distinguish without seeing them side by side, though even then specialists have been known to confuse them. It has often been claimed that false coral snakes are mimics of very venomous true coral snakes. This interpretation presents two problems: firstly, it is difficult to see how the coral pattern could have been selected for when, as we have seen, predators do not have the chance to learn to avoid it; and, secondly, the less and non-venomous mimics generally outnumber the highly venomous models by as many as four to one in parts of South America, while there are no models at all in south-western regions of North America. However, if Wickler's interpretation is accepted, the moderately venomous species are regarded as the models and both the harmless and the highly venomous species as the mimics. The difficulties then largely disappear, for predators can learn to avoid the coral pattern of the moderately venomous species and models outnumber mimics by about three to one. The moderately venomous species can be regarded as Müllerian mimics of each other, the harmless species as Batesian mimics, and the highly venomous species as Mertensian mimics.

There are a number of other controversial examples of warning colouration and mimicry among snakes. It is, for instance, difficult to explain the highly specific resemblance of the harmless South-east Asian Colubrid snake, *Cylindrophis rufus*, to the highly venomous South-east Asian coral snake, *Maticora intestinalis*. When molested, the harmless species has a display identical to that of the coral snake, turning on its back to expose the black and white banded pattern on its belly. Both species also have a bright red underside to their tail which they display as a 'false head' to deflect the attack of predators from their true head. No similar moderately poisonous species exists to act as a model for both, and one is forced to conclude that the poisonous species is a genuine model. The evolution of the pattern and behaviour, and their true significance, are still unsettled.

Finally, there are several non-poisonous snakes which protect themselves with nauseous secretions from glands in the cloaca. The European grass snake, for example, produces a foul-smelling milky fluid when it is alarmed, combining this defence with a display in which it 'shams dead', exposing

The highly poisonous coral snake (above) and the harmless milk snake are probably mimics of similarly coloured species which, being numerous and only slightly poisonous, predators can learn to avoid. The coral snake is a Mertensian mimic of the slightly poisonous species; the milk snake is a Batesian mimic.

its conspicuous pale yellow belly. The brightly coloured garter snakes of North America secrete a similarly nauseous fluid. These secretions serve the same purpose as those of skunks and many insects.

SECOND LINES OF DEFENCE INVOLVING BLUFF AND FRIGHTENING DISPLAYS

Many small cryptic animals, particularly insects, have a second line of defence which comes into operation if their camouflage or disguise is penetrated by a predator. The second line of defence usually takes the form of a frightening display, involving the exposure of brightly coloured false eyes, threatening movements and sometimes hissing. As a result, an animal apparently transforms itself into a frightening object. Such displays are known as Protean displays after the legendary Greek god Proteus, who used to escape from his enemies by changing his form. These displays are a bluff, for the species concerned are harmless and palatable. They work because they exploit the instinctive fear most predators, particularly birds, feel for snakes and large staring eyes. Their instinctive fear of snakes has already been explained, while their fear

of staring eyes probably stems from the resemblance the eyes show to those of their own predators, notably hawks, owls, cats and other predatory mammals. The resemblance is all the more frightening because the false eyes usually look as though they are focused, like those of a predator about to pounce. Staring eyes are a very aggressive stimulus, even to man—perhaps a manifestation of our own latent fear of predators.

The displays used as a second line of defence by chameleons and many lizards involve loud hissing or puffing and the display of the coloured inside of their mouth. The effect emulates the threat display of a viper, and the resemblance is enhanced by lunging movements like those of a striking snake. Cott graphically described the display of the East African species, *Chamaeleo dilepis*:
'When alarmed or angered, they can with startling suddenness undergo transfiguration: the green dress becomes a black one; the animal swells to twice its natural size; it exhibits the bright interior of its mouth; it hisses like a snake. Self-effacement has been replaced as a policy by self-advertisement and threat.'
Cott went on to quote a description of an encounter between a chameleon and a fox terrier:

FOGDEN

'The chameleon invariably tried to run away when attacked, but those who know the species can imagine the ludicrous ineffectiveness of a chameleon's flight. In a few seconds the impossibility of escape seemed to reach the animal's brain, when it at once turned round, opened its great pink mouth in the face of the advancing foe, at the same time rapidly changing colour, becoming almost black. This ruse succeeded every time, the dog turning off at once.'

This display terrifies many laymen and is responsible for the widespread belief, held throughout the Mediterranean region, Middle East and Africa, that chameleons are deadly poisonous. Africans living in the Impenetrable Forest, in Uganda, brought Jackson's horned chameleons to us on the end of ten-foot-long branches and could on no account be induced to hold one.

Among other vertebrate examples of snake-mimicry, there are several hole-nesting birds, notably tits, which mimic snakes by hissing when they are disturbed at their nest. Anyone who has been startled by the explosive hiss of a blue tit, coming from a dark cavity in a tree, will be aware of the irrational fear that it temporarily induces; one is not at all inclined to put a hand into the nest.

By far the most remarkable examples of snake mimicry are found among hawkmoth caterpillars. There are species throughout the tropics with exceedingly realistic false eyes, reptilian scale patterns, and even, in one species, the appearance of a forked snake-like tongue. L. P. Brower described the display of one tropical American species:

'The behaviour and colouration of ... *Leucorhampha* is perhaps the most remarkable Protean display in the entire animal kingdom. About four inches long, this larva lives upside-down on vines along the edge of the forest. It is very difficult to see from a distance, but when its vine is jiggled, the larva lets go with all but its hind legs, swells up its anterior

Far left: The caterpillar of the Malaysian hawk moth, *Rhyncholaba*, has a frightening display in which it expands its anterior segments, exposing a pair of enormous false eyes, and lashes from side to side in a snake-like manner.

Left: As a defence against birds, the African flower mantis, *Pseudocreobotra wahlbergi*, suddenly displays a pair of apparently focused, predator-like, false eyes.

The caterpillar of the Malaysian hawk moth, *Panacra sp.*, is avoided by birds on account of its snake-like appearance.

to expose previously hidden but now most realistic eyes, and immediately starts swaying in a sinuous and snake-like fashion. When I touched the posterior end of one of these larvae, it repeatedly lashed its head back and forth in a striking motion characteristic of Viperid snakes.'

Again, there is evidence that this form of snake mimicry is frightening to many caterpillar-eating animals, particularly to birds.

False eyes are also widespread among butterflies and moths and there is at least one superb example among the praying mantids. The false eyes of most species are on their hindwings, where they are normally concealed by cryptically coloured forewings. Their effectiveness as a defence against predators depends largely on the abruptness with which they are revealed when the insect is about to be attacked. However, there are some species in which false eyes are permanently on display. Predators probably see these conspicuous eyes from a distance and shy away from them without ever noticing their otherwise cryptic owner.

The effectiveness of false eyes as a defence against Passerine birds has been investigated experimentally by D. Blest, who duplicated the way in which many insects reveal their false eyes to birds by suddenly projecting images of eye-models alongside mealworms that birds were about to eat. By varying their appearance, he was able to show that realistic eye-models, apparently focused and complete with highlights, are most frightening to birds and most effective in deterring them from eating mealworms. Blest also experimented with live butterflies. He showed convincingly that birds are frightened by the sudden display of false eyes by peacock butterflies and other species, and also that they cease to be afraid when the insects are rendered 'eye-less' by having the coloured scales removed from their wings. However, it should not be thought that all eye-spots have an intimidating effect. Some small eye-spots on insects have a deflective function, described later, and some on birds and fishes have a social function.

A very large number of cryptic animals display vivid patches of colour, rather than false eyes, as a second line of defence. Most are mantids, grasshoppers, stick-insects, bugs, moths and other insects, but a few are frogs and lizards. Some of these animals are genuinely distasteful and their displays of brilliant colour can be regarded as simple warning colouration. Such species benefit from being cryptic as well as being distasteful.

Other species are perfectly edible and the conspicuous colour patterns they display can be regarded as Batesian mimicry. However, this mimicry seldom involves a specific model, and must depend for its effect upon predators showing a generalized avoidance of black and red, black and yellow, or whatever warning pattern is appropriate.

Some of these animals with concealed patches of colour on their wings or flying membranes—notably grasshoppers, stick-insects and flying lizards—differ from the rest by displaying them only in flight. They are said to display flash colouration. It has been suggested that when these animals alight, the sudden disappearance of colour, together with the sudden cessation of movement, tends to confuse the eyes of pursuing predators. However, many naturalists who have watched these animals in the field find this interpretation difficult to accept, for their flashing colours seem to make them easier to see and to follow than would otherwise be the case. A more likely explanation is that flash colouration has a social significance. Certainly the flash colouration of many grasshoppers is very prominently displayed in aerial song-flights during their breeding seasons. It is also significant that closely related species of grasshoppers, stick-insects or flying lizards that occur together in the same environment can be distinguished easily by the colour of the patches they display. For example, four species of flying lizards living in rainforest near Kuching, Sarawak, have flying membranes coloured mainly brick-red, dark brown, yellow or blue, each colour being characteristic of a particular species: that is, the colours probably act as species recognition signals.

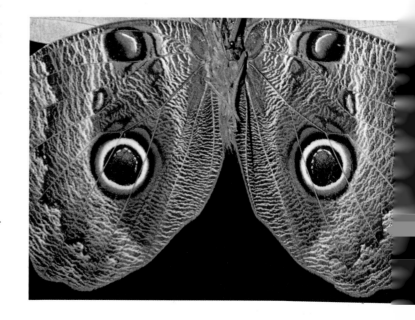

FOGDEN/COLEMAN INC

Below, left: The owl butterfly gets its name from the fancied resemblance of the underside of a set specimen to the face of an owl. In reality it rests in a typical butterfly position, so that only one false eye is ever visible, though even this may have an intimidating effect on predators. The eyes reach a high degree of perfection, even having simulated highlights.

As a first line of defence, this distasteful lantern bug from Borneo is cryptic, the colour of its forewings matching the mossy tree-trunks on which it rests. If it is detected, it has a second line of defence: it exposes its warningly coloured red and black hind-wings and waxy, white abdomen.

If its false eyes are exposed, the peacock butterfly is usually avoided by birds, but it is readily predated if the scales forming the eyes are experimentally removed.

As a second line of defence, swallowtail caterpillars evert a brightly coloured forked organ, called an osmeterium, which disperses a repellent odour.

BEAMES/ARDEA

PAUL POPPER

Special deflecting, alluring and mimetic colouration

Matching the colour of their chosen flower to perfection, many crab spiders sit in ambush for butterflies and other insects that are attracted by nectar. In taking advantage of flowers in this way, such crab spiders are examples of adventitious or acquired alluring colouration.

In this chapter are grouped a wide range of different colour adaptations intended to deceive or attract other animals. Some, like deflection marks, have a protective function, but others are aggressive adaptations which lure animals into striking distance of predators, or result in their being exploited in some other way.

COLOURS AND STRUCTURES WHICH DEFLECT THE ATTACK OF PREDATORS

Deflection marks provide yet another example of colour adaptations that are intended to deceive and confuse predators. They are particularly common among insects, often taking the form of small eyespots and other structures, which give the impression that the insect's head is at the posterior end of its body. It has been shown experimentally that predators preferentially direct their attack at the heads of insects or at any small, conspicuous spots that look like eyes, presumably because insects are most easily killed or incapacitated by damage to their head. Strategically placed eye-spots or head-like structures therefore deflect the attack of predators to the least vulnerable area.

False heads and eye-spots interfere with the attacking techniques of birds in another way, for birds anticipate that their prey will try to escape in the direction in which they are facing and make appropriate allowances. By appearing to face the opposite direction to that in which they actually do fly off, the insects confuse predators and cause them to misjudge their attack.

Eye-spots and structures resembling heads are particularly common on the hindwings of butterflies. If birds manage to get hold of one of these butterflies, they generally end up with no more than a beak-full of torn wing and the butterfly is otherwise unharmed. The numerous butterflies that fly around with beak-marks on their wings are witness to the success of this deception. C. F. M. Swynnerton investigated the effectiveness of deflection marks on a species of *Charaxes* butterfly which does not normally have them. He painted small conspicuous eye-spots onto the underside of the hindwings of 51 specimens and then released them. Subsequently, 47 of these specimens were recaptured with injuries to their wings, most of which appeared to be due to the attacks of birds directed at the eye-spots. The injuries involved both hindwings, showing that the butterflies had been attacked while settled, with their wings

FOGDEN/COLEMAN INC.

closed in the normal cryptic position. Swynnerton also found evidence that the artificial eye-spots helped the individuals on which they had been painted to survive longer than the individuals without them.

False heads reach their greatest degree of elaboration in Lycaenid butterflies, many of which have 'tails' on their hindwings which resemble legs and antennae. The butterflies draw attention to these decoys by moving their hind-wings. A few species also turn around immediately after alighting so that their false head faces the direction they were flying in. Sometimes they even move a few steps backwards, further deceiving any watching predator.

Fishes are another group in which eye-spots are very common, although their significance is by no means always clear. However, there are marine reef fishes, particularly in the genus *Chaetodon*, in which posterior eye-spots are combined with eye camouflage. Wickler has evidence that these eye-spots, at least, deflect the attacks of sabre-toothed blennies, small fish that dart forward from concealed positions to attack the soft skin around the eyes of larger fish. It is also possible that they lead large predatory fish into misjudging their attack, while their owner darts safely away in an unexpected direction.

Deflection marks are not uncommon among vertebrates such as snakes and lizards. The Malaysian coral snake, *Maticora intestinalis*, and its mimic, *Cylindrophis rufus*, wriggle their tails in the air when they are alarmed, thereby exposing the brilliant red undersurface. Similar adaptations and displays are found in other snakes in other parts of the world. It is significant that several, from widely separated areas, are known locally as two-headed snakes. Their brightly coloured and wriggling tail-tip is by far their most conspicuous feature, and almost certainly diverts the attack of predators away from their vulnerable head. It would, of course, be dangerous for any predator that feeds on snakes to get a firm grip on the tail, rather than

Left: Like many cryptic butterflies, this Malaysian jungle glory has small eye-spots on its hindwings which improve its chances of survival if its camouflage fails. Unlike large false eyes, small eye-spots stimulate pecking and so deflect bird attacks from the butterfly's vulnerable body.

Like many other Lycaenid butterflies, the Malaysian back-to-front butterfly, *Zeltus amasa*, has false eyes, antennae and legs on its hindwings, which divert attention from its less conspicuous genuine head. When attacked, the butterfly darts off in an un-expected direction, leaving the predator with a fragment of wing

FOGDEN

The South-east Asian lantern bug, *Ancyra annamensis*, has a very convincing false head which causes predators to direct their attack to the wrong part of its body.

Right: The colourful tail of the Californian blue-tailed skink breaks off easily, distracting the attention of predators while the lizard makes good its escape.

the head, for if the snake in question was poisonous they would run a grave risk of being bitten and killed. In the face of this possible danger, it is likely that the appearance of a 'two-headed snake' is sufficiently alarming to deter many predators.

The deflection marks of lizards are rather different. It is well known that many species, particularly skinks and geckos, have fragile tails that break off if seized by a predator. Once broken off, the severed tail wriggles vigorously, diverting the predator's attention while the lizard makes its escape. Many of the diurnal species have this dispensible part of their body so brightly coloured that predators actually attack it preferentially, thus ensuring that the lizards have a good chance of escaping. The tail regenerates quickly and the large number of individuals with replacement tails

is a good indication of the value of this adaptation.

Finally, mention must be made of the way in which many ground-nesting birds deflect attention from their vulnerable eggs and young. In many ducks and gamebirds the brightly coloured and conspicuous male has nothing to do with incubation or rearing the young, but helps indirectly by drawing the attention of predators away from the nest and the cryptic female. Other ground-nesting birds have special distraction displays, appropriately called broken-wing displays, by which they induce predators to follow them. They encourage pursuit by giving a realistic imitation of an injured bird; they flutter along the ground, making themselves as conspicuous as possible by dragging their wings, fanning their tail and exhibiting contrasting patches of colour on their plumage.

BURTON/COLEMAN

Its real eye concealed by a disruptive black band, the butterfly fish, *Chaetodon auriga*, has a false eye-spot near its tail. This diverts the attention of the sabre-toothed blenny, *Runula*, a small fish that preferentially feeds on skin around the eyes of larger fish.

Right: The flower-like appearance of the Malaysian mantis, *Hymenopus coronatus*, lures nectar-seeking insects to their death.

This Limacodid slug caterpillar from Borneo has poisonous spines and is avoided by experienced predators. The black markings on its hind-end help it to survive the attacks of inexperienced predators by deflecting them from its vulnerable head.

FOGDEN/COLEMAN INC

FOGDEN

ALLURING COLOURATION

Just as colours and structures can be used by animals to deflect predators, so they can be used by predators to attract their prey. A particular part of the animal may be coloured and modified to act as a lure, or the whole animal may be attractive. Alluring colouration occurs only in cryptically coloured animals that ambush their prey and their ability positively to attract victims to them represents a considerable advance over the ambush techniques of most cryptically coloured animals, which rely largely on chance to bring their victims close enough to be caught.

The best examples of animals whose whole body is adapted to act as a lure are a praying mantid and a spider. The Malaysian mantis, *Hymenopus coronatus*, is modified to resemble a flower and preys on insects attracted to it. It varies in colour from pink to white and matches the flowers among which it almost invariably sits. It resembles a flower so closely that it is attractive to insects even when it is resting alone among green leaves.

Even more remarkable is a Malaysian crab spider that resembles a bird-dropping and preys upon butterflies and other insects that are attracted

by the moisture and salts in bird excrement. The spider's body simulates the black-streaked, whitish, solid part of the dropping, while a thin film of web, which it spins around itself, simulates the white liquid part, which spreads and dries into a thin encrustation. The web even has a drop-like extension which runs down the slope of the leaf, in exactly the same way as a genuine liquid film would do. The spider sits on the web, gripping it with its two short hind pairs of legs, while its much larger front legs are held folded in front of it, ready to seize its prey. This spider feeds by night as well as by day and presumably has an alluring smell as well as an alluring appearance, for it is unlikely that many nocturnal moths would be attracted to it otherwise. Under cover of darkness it adopts a different position, more suited to grasping prey quickly. Its front two pairs of legs are held out sideways, like the open jaws of a gin-trap, and spring together to grasp its victims. Such a position could not be used by day, for it would spoil the disguise and might attract birds.

Animals that use special structures to lure their prey are relatively numerous, particularly among aquatic animals. The superbly camouflaged angler-fishes, for example, attract their prey by a variety of coloured worm-like lures, some of which are modified dorsal fin-rays, others modified processes of the lower jaw. The alluring apparatus of *Lophius piscatorius*, for example, is a modification of the first dorsal fin-ray. It is situated well forward on the snout, is elongated, flexible and ends in a flattened worm-like appendage, which waves in the water and attracts the attention of the small inshore fishes upon which the angler feeds. Once its prey is close by, the angler lunges forward and engulfs it.

Other aquatic animals that use coloured worm-like lures include a freshwater catfish, *Chaca chaca*, and the South American freshwater mata-mata turtle, both of which have wriggling red 'worms' at the corners of their mouths, and the North American alligator snapping-turtle, which has a red or white grub-like process on its tongue. Such devices are not entirely confined to aquatic animals, for the young of certain Crotaline vipers are said to attract prey by wriggling the brilliant yellow worm-like tip of their tail and South American horned frogs wriggle one of the fingers of their hands.

There may also be examples of alluring colouration among birds. The American king-bird, for example, has been observed catching insects which

FOGDEN

This crab spider from Borneo, which resembles a bird-dropping, is a fine example of alluring colouration. It preys on butterflies and other insects that are attracted by the salts and other constituents of genuine droppings.

Right: The angler-fish, *Antennarius*, preys on small fishes attracted by the worm-like lure on its snout.

had apparently mistaken its orange-red crown for a flower. It is conceivable that the enormous bright yellow gape of the basically nocturnal frogmouths is attractive in a similar way, for they have been seen sitting by day with their bills wide open.

A rather different form of alluring colouration is found in several predatory invertebrates that take advantage of flowers to attract their prey. A number of praying mantids and crab spiders, for example, habitually sit in or among flowers, match them in colour, and can be said to have adventitious alluring colouration. Some species are able to vary their colour to match different flowers, which is a great advantage in areas where different plants have short flowering seasons. The same adventitious alluring principle is used by lions, leopards and other large predators, which lie in wait for their prey at water-holes during the dry season.

THE CLEANER-WRASSE AND ITS MIMIC

The term cleaner is applied to a number of marine fishes and shrimps that live entirely, or partly, by removing and eating ectoparasites, diseased tissue and fungus from the skin of other fishes. The cleaning is carried out at special stations, each tended by a pair of cleaners. I. Eibl-Eibesfeldt has described the diversity of fish that assemble at these stations, often coming from a considerable distance:

'Many different species of fish allow themselves to be cleared by cleaners: predatory fish and peaceful ones, reef-dwellers as well as fish of the open seas. Sometimes we saw swarms of fish appear out of the deep blue water and stand head-down above a cleaner station as if in response to a single command. They waited for the cleaners, which soon were busy with their task. After several minutes the fish school again disappeared in the depths of the sea.'

Some idea of the immense importance of cleaners in tropical seas is provided by an experi-

ment described by C. Limbaugh, in which all known cleaners were removed from two small isolated reefs with an especially abundant fauna. Within a few days the number of fish had decreased dramatically, and after two weeks there remained only a number of territorial fish, which bore numerous fungal patches, loose fin fragments and skin abscesses. Only when new cleaners had colonized the area did the fish population increase to its previous level. Limbaugh also observed that a single cleaner was visited by as many as 300 customers of various species within a period of six hours.

The relationship between cleaners and their customers is a mutually beneficial partnership, in which the cleaners are provided with a food source, and their customers a service which keeps them healthy. An elaborate system of signals has evolved between the cleaners and their customers. Various cleaner shrimps in the genera *Stenopus* and *Periclimenes*, for example, attract the attention of their customers by waving their long antennae, and climb onto their surface when they exhibit open-mouthed inviting signals. Similarly, the much studied cleaner-wrasse, *Labroides dimidiatus*, has characteristic colouring and a special 'cleaner dance' as it approaches its customer, thereby ensuring that it is recognized and not eaten by the more voracious of them. The cleaner-wrasse taps its customer with its vibrating ventral fins to keep it informed as to the area it is cleaning, and butts at the mouth or gills when it wishes to enter to clean inside. The customer responds accordingly by stopping the movement of fins that are being cleaned and by opening its mouth and gill-covers to give easy access. Some customers even change colour in order to provide a contrasting background against which parasites are more easily visible. *Naso tapeinosoma*, for example, turns pale blue when it is being cleaned. When they wish to breathe deeply, the customers signal for the cleaner-fish to leave their mouth by jerking it half-closed and opening it again, and they signal their intention to swim away by shaking their entire body.

In the vicinity of cleaning stations manned by *Labroides dimidiatus*, there are sometimes cleaner mimics which exploit the non-aggressive behaviour of large fish that are being cleaned. These mimics, which are a species of sabre-toothed blenny, have been studied in great detail by Wickler:

'The fish resembling the cleaner is *Aspidontus taeniatus*. It is exactly similar in size, colouration, and swimming behaviour, and even exhibits the same dance as the cleaner. Fish which have had experience with the cleaner will also position themselves unsuspectingly in front of the mimic, showing the invitation posture for cleaning described above. They then receive a nasty surprise. The mimic approaches carefully and bites off a semi-circular piece of the victim's fin and eats it.

CASSELLI (AFTER KACHER)

Taking advantage of its resemblance to the cleaner-wrasse, the sabre-toothed blenny, *Aspidontus*, approaches fishes that expect to be cleaned and bites pieces out of their fins and tail. Here the cleaner-wrasse is above, its mimic below.

Right: By having distinctive colouring and swimming in a special way, the cleaner-wrasse, here cleaning a Maori cod, advertises its services and avoids being eaten by large predatory clients.

The African flower mantis, *Pseudocreobotra wahlbergi*, sits in ambush for nectar-seeking butterflies and other insects, using flowers as a lure.

FOGDEN

The fish immediately jerks round after the jab, but the mock cleaner calmly stays put as if knowing nothing about it, and remains unmolested because of its cleaner's costume. Genuine cleaners, of course, sometimes nip the customer, and cleaning shrimps may even make small incisions in the skin with their chelae in order to extract embedded parasites. But the bite of the mimic *Aspidontus*, judging from the reaction of the victim is much more painful. Fish which have been repeatedly bitten in this way become distrustful even towards genuine cleaners.'

The survival of the cleaner mimic depends on its very close resemblance to the cleaner-wrasse, and the way in which both species correspond in colour from area to area Wickler has described: '*Aspidontus* is in fact one of the most exact mimics we know. The model *Labroides dimidiatus* occurs as a number of local races within the area of distribution, with corresponding differences in colouration. Here and there, the cleaner is characterized by a small or large black vertical stripe at the base of the pectoral fins, while the cleaner population near Makatea in the Tuamotu archipelago bears an orange-red spot on the flanks. In every case, the local mimic population shows the same special colouration and is thus similar to the model in all areas.'

MIMICRY IN PARASITIC BIRDS

Another form of aggressive mimicry is practised by such parasitic birds as whydahs and cuckoos, whose eggs and young are cared for by members of another species.

The African Viduine whydahs lay their eggs in the nests of Estrildine finches, to which they are quite closely related. A female whydah usually lays only one egg in any one nest of its host and does not destroy any of the host's eggs. However, several eggs are sometimes laid in the same nest by different female whydahs, and when this happens the late-comers are said to destroy some of the host's eggs, leaving those of their own species unharmed. Apparently they are able to distinguish Viduine eggs more easily than their host can, for the host is not known to reject them. Both types of eggs are white, but whydahs' eggs are usually larger and more oval in shape.

The young whydahs are reared with the host's young, which they closely resemble, though adults of the two species are strikingly different. The young of Estrildine finches have characteristic gape-markings which stimulate feeding by the adults; interestingly, the parasitic young have gape-markings which stimulate feeding by the adults; interestingly, the parasitic young have gape-markings which are almost identical to those of their normal host. They also mimic the begging calls and peculiar begging postures of Estrildine young. Without such markings and behaviour, it is doubtful whether the parasitic young would be fed by the host. In fact, it seems likely that Estrildine species have evolved characteristic gape-patterns in order to counteract nest-parasitism; if so, whydahs have kept pace and simultaneously evolved similar patterns. Because of the need for a specific resemblance to one host, each species of whydah normally parasitizes only one Estrildine species, though the system sometimes breaks down in areas where a particular host is rare or absent. In parts of western Uganda where its normal host, the common waxbill, is very scarce, the pin-tailed whydah parasitizes different Estrildines; these are mainly red-billed firefinches, which are not normally parasitized by whydahs. They have less complicated gape markings and may well be less discriminating than the normal host.

Another interesting feature of Viduine mimicry is that the male's song has many phrases identical to those of its usual host. The male probably learns these song phrases while it is a nestling and they may help to ensure the selection of the appropriate host. However, the full significance of this song mimicry is complicated and has not been fully explained.

The system of parasitism and mimicry practised by the crested cuckoos of the genus *Clamator* is rather similar to that of the whydahs, in that the young cuckoos are reared together with the host's young. The female cuckoo substitutes one of its own eggs for an egg of the host, usually a crow or a babbler, so ensuring that the clutch remains the optimum size for the host to rear successfully. Both egg and young closely resemble those of the host. The system practised by the European cuckoo and other species in the genus *Cuculus* is different. The adults are much bigger than the usual host species and the hosts are quite unable to rear a young cuckoo together with their own young. For this reason, the young cuckoo ejects the unhatched eggs or newly hatched young of the host from the nest as soon as it hatches, thereafter appropriating all food brought by its foster parents. The foster parents accept the young cuckoo, even though it bears no resemblance to their own young, which are not, of course, there for comparison. Any

The long-tailed paradise whydah lays its eggs in the nests of the green-winged pytilia. Because the pytilia feeds nestlings only if they have the right gape pattern, the whydah nestlings (left) have evolved gape markings that mimic those of its host's nestlings.

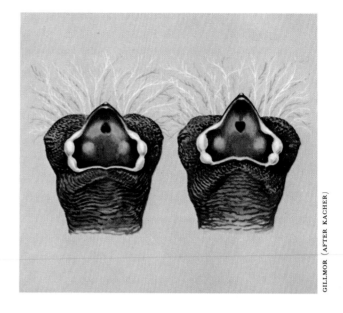

Eggs of the European cuckoo and some of its hosts: top left, meadow pipit; top right, great reed warbler; bottom left, grass-hopper warbler; and bottom right, redstart. In each pair of eggs, that of the cuckoo is on the right. Egg size as well as colour is involved in this mimicry, for the cuckoo eggs are exceptionally small for a bird of its size.

135

brightly coloured gape stimulates feeding in most Passerine birds, and the enormous gape of a young cuckoo is a 'super feeding stimulus'. The begging of a young cuckoo is in fact so stimulating that other birds passing by with food sometimes cannot resist feeding it at the expense of their own young.

The eggs of the European cuckoo generally mimic those of their host very closely. Many different hosts are parasitized, but each individual female cuckoo lays only one type of egg, and almost always parasitizes the species by which it was itself reared; thus female cuckoos with eggs like those of reed warblers were probably reared by reed warblers, and generally parasitize reed warblers themselves. However, the degree of similarity between the eggs of the European cuckoo and its hosts' varies considerably, depending perhaps on the length of time the cuckoo has been parasitizing a particular species. As time passes, there is selection in favour of the hosts that most successfully reject cuckoo eggs, and in favour of the cuckoos whose eggs are rejected least often, resulting in an increasingly better match between their eggs. Eventually, the host species may become so discriminating that cuckoos are more successful when they parasitize a different host, even if their eggs are not a good match. It is possible that cuckoos have only recently started to parasitize dunnocks in Britain, for their eggs are very poorly matched, those of the cuckoo being speckled and variable in ground colour, while those of the dunnock are a clear bright blue. 'Dunnock cuckoos' may well evolve blue eggs in time, just as cuckoos elsewhere in Europe have evolved blue eggs to match those of redstarts.

An interesting feature of adult cuckoos in the genus *Cuculus* is that they closely resemble various species of sparrow-hawk, while the Asian drongo-cuckoo resembles a drongo. It has been suggested that the function of these mimetic resemblances is to lure the host species away from their nest; the subject is controversial and needs further study.

COLOUR SIGNALS FROM PLANTS
TO ANIMALS

Many plants have evolved protective and aggressive adaptations for their relationships with animals. There are species that protect themselves from herbivorous animals with thorns, spines and irritating hairs and others that are poisonous; some cacti are camouflaged to resemble stones; and there are even plants, such as sundews, butter-worts, venus' fly-traps and pitcher-plants, that are semi-carnivorous and lure insects into traps. The most striking and interesting relationships between plants and animals are mutually beneficial partnerships, in which plants supply animals with an attractive source of food in return for the services of pollination and seed-dispersal.

What people think of as flowers are structures which have evolved specifically to attract the insects, birds or bats, that bring about their pollination by carrying pollen from one flower to another. Flowers attract these animals with nectar, pollen or edible petals and use colour and scent signals to advertise themselves. The food they supply, and their appearance, depends entirely on the needs, behaviour and sensory organs of the animals that bring about their pollination. Flowers that attract insects provide relatively small quantities of nectar, while those that attract birds provide much more. Some species of bird-pollinated *Erythrina* drip so much nectar from their flowers that they are called 'cry-baby trees' in the West Indies. Flowers that attract bats also provide large quantities of nectar and some have edible and nutritious petals. Other plants, such as *Cassia* species, have even developed nutritious sterile pollen as an attractive food source for bees.

Flowers that are pollinated by diurnal insects, particularly bees, advertise themselves by being blue or yellow; they are seldom red, for bees and most other insects are insensitive to red light. Many flowers also reflect ultra-violet light and sometimes have superimposed ultra-violet patterns which are visible to insects but invisible to man. Cinquefoils and evening primroses, for example, reflect ultra-violet light over most of their area, but not from a patch at their centre, which appears as a contrasting patch to bees and other insects. Such patches are called nectar-guides because they guide insects to the nectaries. Poppies, being red, are an apparent exception among insect-pollinated flowers, but they too reflect ultra-violet light and therefore appear ultra-violet, not red, to insects. By contrast, flowers that are pollinated by birds almost invariably advertise themselves by being red because birds are particularly sensitive to red light, while flowers pollinated by nocturnal insects or bats are white, so that they are visible in dim light. To them colour signals are less important than scent signals and at night they release strong scents which attract their pollinators from a considerable distance. Scents that attract bats are particularly strong and are often reminiscent of

An elephant hawkmoth on white campion. Flowers that are pollinated by night-flying moths provide nectar and open in the evening, attracting the attention of their pollinators with a fragrant scent and with white petals that are visible in dim light.

rotting fruit. A Malaysian bat-pollinated tree, *Oroxylon*, is known as 'the midnight horror', because its flowers open and emit a dreadful stench in the middle of the night. Flowers pollinated by diurnal insects also have scents, and special odour-guides, distinct from the general odour of the flower, are sometimes associated with the coloured nectar-guides. Scents are almost completely absent in large showy red flowers that are pollinated by birds, as, with rare exceptions, birds have no sense of smell.

Flowers are adapted to suit their pollinators in other ways: the petals or flower-heads of insect-pollinated flowers provide alighting platforms for bees, butterflies and other insects and some bird-pollinated flowers have suitable perches for birds just below them. Other bird-pollinated flowers

BURN

RIVAROLA/COLEMAN

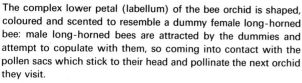

The complex lower petal (labellum) of the bee orchid is shaped, coloured and scented to resemble a dummy female long-horned bee: male long-horned bees are attracted by the dummies and attempt to copulate with them, so coming into contact with the pollen sacs which stick to their head and pollinate the next orchid they visit.

Bird-pollinated flowers are appropriately adapted, having red or orange petals, little or no scent and a copious supply of nectar. New World species, like this one, are pollinated mainly by hummingbirds, Old World species mainly by sunbirds.

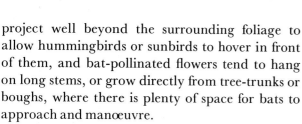

Right: Many plants have evolved red or orange fleshy fruits to attract fruit-eating birds, such as the waxwing. The hard seeds pass unharmed through the bird's gut and are dispersed.

project well beyond the surrounding foliage to allow hummingbirds or sunbirds to hover in front of them, and bat-pollinated flowers tend to hang on long stems, or grow directly from tree-trunks or boughs, where there is plenty of space for bats to approach and manœuvre.

Pollination by animals is more reliable than pollination by wind or water, but there is still a need for plant adaptations to ensure that the animals move between flowers of the same species, rather than to several different species. Hence, flowers compete with each other to attract and retain the attention of pollinators and a major reason for the diversity of their colours is the need for each kind of flower to be recognizable to pollinators. Some plants have evolved adaptations which ensure that their flowers are visited by only one species of pollinator. The most extreme examples are found among certain orchids, whose

flowers mimic in their structure and colour the females of certain bees, wasps or other Hymenoptera. The males of the appropriate Hymenoperan are attracted to the dummy females and are stimulated to copulate with them. The orchid is so constructed that its pollen sacs come into contact with the copulating insect and adhere to it; later they are transported to another nearby orchid and bring about its pollination.

Colour signals are also involved in the dispersal of seeds, for the red colour of many fruits is an advertisement to attract fruit-eating birds. Birds eat the fruit and disperse the undigested seeds in their excreta, often after they have flown a considerable distance from the parent plant. The plants concerned have particularly bountiful crops of fruit, which ripen progressively over a long period of time, thus increasing the effectiveness of this mutually advantageous arrangement.

social uses of colour

A few exceptional animals live their lives with virtually no social contact with other members of their species, even for breeding. The males of some terrestrial arthropods, for example, deposit spermatophores more or less at random for the females to find eventually by chance. However, most animals meet other members of their species at some stage in their lives, if only to breed, while many live in social groups throughout the year, co-operating in their search for food and finding safety in numbers. Such animals have to be able to recognize and communicate with each other and to do this they use signals involving their best developed senses.

The use of colour signals for social communication is virtually confined to animals that are active by day and have good colour vision. A great

Previous page: The brightly coloured gape of nestling dunnocks and other birds is a signal that stimulates parent birds to feed them.

Right: This robin is attacking a stuffed robin that has 'intruded' into its territory. It would even attack a tuft of red feathers, though not a stuffed juvenile robin, which has no red breast. This proves that the red breast is the stimulus that releases attacking behaviour. Male robins even attack their potential mate until a strong pair bond has been formed.

Male yellow-shafted flickers recognize rival males by their black moustachial streaks and attack even females if black streaks are experimentally painted on them.

The bright red throat and belly of male three-spined sticklebacks is assumed only for breeding, for it increases their vulnerability to predators. Their red belly is a social signal that is attractive to females and threatening to rival males.

Above, right: Male and female blue-footed boobies have bright blue feet which they display prominently to each other in their courtship displays.

Below, right: In the red-necked phalarope there is a reversal of sexual roles: the female is the more brightly coloured of the sexes and courts the male, while the male cares for the eggs and young.

contrast a realistic stuffed juvenile robin without red feathers was totally ignored. Similarly, a classic experiment by N. Tinbergen showed that the red belly of a male three-spined stickleback releases attacking behaviour in other males. Even the crudest red-painted fish models released more aggressive responses than a realistic stickleback model without a red belly. Tinbergen also showed that the red colour must be on the underside of the fish, proving that, in this case, the releaser has two essential qualities—being red and being underneath.

Though releasers are particularly important in social signalling, they are not confined to interactions between individuals of the same species: the colours of flowers release responses in birds and insects and the protective false eyes of moths and caterpillars release alarm responses in insect-eating birds. It must also be emphasized that releasers can involve the senses of smell, hearing, taste and touch, as well as sight.

Simple colour releasers are commonly used by animals to enable them to recognize members of their own species, the opposite sex, sexual rivals and even their young. Adult zebra finches, for example, recognize juveniles by their black bill and do not feed them, in spite of their intense begging, if their bills are painted red. Male yellow-shafted flickers recognize rival males by their black moustachial streaks and will attack even their own mate if black streaks are painted on her. Similarly, male fence lizards recognize rivals by the blue patch on their throat and the blue streak along the edge of their belly: they attack females that are painted with similar blue markings and court males whose markings have been covered over.

Colour releasers are particularly important in the territorial and courtship displays of visual animals, their effect often being greatly enhanced by exaggerated and stereotyped movements and postures. An aggressive robin, for example, turns its red breast directly towards a rival, stretches upwards to display the greatest possible area of red and sways from side to side; threatening male sticklebacks show as much as possible of their red belly by displaying in a vertical head-down posture; and blue-footed boobies display their bright blue feet to advantage by waggling them as they come in to land on their territory and by strutting around with an exaggerated high-stepping action.

In most species the individuals with the brightest colours are the most successful in social interactions

deal has been written about this type of signal, largely because man himself is a visual animal and is well attuned to perceive and interpret the visual signals of other species. In fact, man's studies of animal behaviour have been very much biased towards the colourful territorial and courtship signalling of such visual animals as primates, birds, lizards, fishes and butterflies.

Most of the colour signals that are used by animals for social communication are known as 'releasers' because they elicit, or 'release', a specific instinctive response from the animal that receives the signal. The European robin, for example, always responds aggressively to any rival male that intrudes into its territory. By presenting a bundle of red feathers to a territorial male robin, D. Lack was able to prove that the red breast alone is the releasing stimulus; the feathers provoked intense aggressive behaviour, while in

and, if they live in groups, are at the top of dominance hierarchies. Brightly coloured males are usually dominant over dull females, though in phalaropes, button-quails and other species in which there is a reversal of the sexual roles, females are more brightly coloured than males. P. R. Marler illustrated the importance of colour in social hierarchies by colouring the undersides of female chaffinches pink to resemble those of males; low-ranking females coloured in this way subsequently dominated all other females.

Though colour signals are obviously valuable for social communication, their use is often restricted by an animal's need to be inconspicuous for other reasons. Hence, releasers involving colours are most prominent in animals that have no need to be cryptic in order to hunt, avoid predators or rear

their young in safety. Parrots can afford to be brilliantly coloured because they feed on fruit, rely on flight to avoid predators and breed in the safety of tree-holes where their vivid plumage is safely obscured. Similarly, many reef fishes can afford to be permanently conspicuous and readily recognizable because they feed on sedentary animals and algae and can retreat to the safety of coral crevices to escape predators and to breed.

Animals that cannot afford to be permanently conspicuous rely either on colour signals that can be displayed temporarily at will, or solely on non-visual signals. Thus many cryptic grasshoppers have special display flights in which they reveal colourful patches on their wings; Cichlid fishes can change rapidly into a conspicuous and colourful dress to threaten rivals; and many cryptic birds

Displaying to a rival—its own image—this male Anole lizard makes itself appear as impressive as possible by standing sideways, flattening its body laterally, expanding its yellow gular pouch and doing rhythmic 'press-ups' on its front legs. Besides being a threat signal to rivals, the display attracts females and maintains reproductive isolation, for the males of each anole species have a characteristic gular pouch colour and 'press-up' rhythm which is attractive only to females of their own species.

Left: In the breeding season, puffins develop brightly coloured, horny sheaths on their bills, which play a part in their threat and courtship displays.

communicate with one another almost entirely by song or other vocalizations.

THREAT AND APPEASEMENT DISPLAYS

Threat displays serve to space out the males of a species by advertising the location and boundaries of their territories. These displays are particularly vigorous between rival males in areas where their territories adjoin and can be considered as a substitute for real fighting, which might be injurious to both parties.

The males of most territorial species have at least one threat display which is obvious at long range; it advertises their presence to distant rivals and reduces the need to patrol their territorial boundaries. The nature of the display depends on

147

the animal's environment. Woodland birds and such forest-living primates as chimpanzees, gibbons and howler monkeys rely almost entirely on vocalizations, while animals living in open country frequently have conspicuous visual displays which are enhanced by colour. The African black-bellied bustard and several northern waders have display flights which show off their white-edged, black bellies; bishop-birds and widow-birds fly back and forth over their grassland territories displaying their black and red plumage or long plumes; Agamid lizards sit prominently on rocks and other vantage points, displaying their bright colours. The white-sided jack-rabbit of Mexico has a particularly effective territorial display. It has special muscles which pull the white skin on its flank and belly over its back, exposing a white area

When displaying to rivals, the male capercaillie (the name is Gaelic for 'old man of the woods') fans its tail, ruffles its neck feathers and utters a curious call, which begins with a rattle and ends with a pop, a gurgle and the noise of its beating wings.

Right: The males of several species of monkeys have a red penis and bright blue scrotum which they display conspicuously when threatening males of rival groups. In the case of the mandrill, the same colour pattern is reproduced on the face and acts as a threat signal in the same way.

which is visible at a considerable distance. E. W. Nelson described its appearance in the field:

'This enlargement of the white area is always on the side turned toward the chance intruder, and accordingly alternates from side to side as the animals zigzag away. In the bright sunlight the snowy white side flashes brilliantly, attracting attention from afar, and affording a fine example of directive colouration.'

The threat displays used by animals at close quarters in territorial boundary disputes are often different from their long distance displays. When their rival is near, they endeavour to appear as impressive as possible by increasing their apparent size and exposing conspicuous patterns and colours. We have already mentioned the way in which territorial robins and sticklebacks contrive to

MARTIN

When foraging on the ground, groups of ring-tailed lemurs probably use their magnificent erect tails as a contact signal. In the breeding season, males also use their tails in stink-fights, marking them with scent from their wrist-glands (visible on the left forearm of this individual) and flicking them at rivals in order to disperse the scent.

Displaying on traditional territorial grounds, male prairie chickens endeavour to appear impressive by fanning their tail, stamping their feet and inflating their yellow wattles as they produce their booming call. The females are attracted to the dominant and most impressive males, which monopolize the centre of the area.

BARTLETT/COLEMAN

display the maximum possible area of red colouration to their rivals. There are countless other examples. Many animals posture sideways at their rivals and enlarge themselves laterally: male bushbucks raise the crest of hair on their backs and erect and fluff their tails like white flags; Anole lizards inflate their bodies and expand their brightly coloured gular pouches and dorsal crests; Cichlid fishes erect their fins and intensify their colours. Other animals face their rivals and enlarge their frontal view; peacocks, sage grouse, black grouse and many related species erect and fan their tails, inflate coloured wattles or expose other patches of colour; ruffs expand their ruffs of patterned feathers and frilled lizards the broad frill around their necks. Male vervet monkeys and baboons sit at the edge of their groups and threaten the males in rival groups by displaying their enlarged and brightly coloured genitals. Some animals draw attention to their weapons: mammals, birds and lizards often threaten with open mouths, showing their teeth or coloured gape; fiddler crabs wave their enormously enlarged and brightly coloured claws.

Though genuine fighting is rare, many male animals test their strength by ritual fighting: antelopes interlock their horns and have fencing, pushing or wrestling matches; fence lizards take it in turns to grasp each other's neck and judge the strength of their opponent's grasp; butterfly fishes fight by head-butting and many Cichlids by beating their tails at each other, creating pressure waves; fiddler crabs grasp each other's claws and try to throw each other over. These ritual fights are usually over quickly, one of the combatants submitting as soon as it realizes it is weaker than its opponent. The loser shows its submission with appeasement displays which rely for their effect on being as different as possible from threats. Most animals try to appear small and insignificant by prostrating themselves, concealing coloured areas, changing to dull colours if they are capable of rapid colour change and turning their weapons away from their opponent. Appeasement displays inhibit further attack and give the loser the opportunity to escape.

COURTSHIP DISPLAYS

The displays used by animals in courtship are basically similar to those used for threat and appeasement; their major function is to break down the barriers which exist between males and females, so that mating can take place. In species in which both sexes care for the young, courtship displays also help to establish and maintain a pair-bond which may last for some time, often for life.

Females are initially attracted to males by the males' long distance threat displays: thus bird song and the visual threat displays of bustards, frigate birds and Agamid lizards attract females at the same time as they repel rival males. A male's first reaction towards a female that intrudes into its territory is usually aggressive, while the female is frightened as well as attracted. The male's aggression is particularly marked in animals in which the sexes look alike, notably such species as grebes, gulls, tits and many reef fishes, but aggression also occurs in species in which the sexes

Among the most unusual courtship displays are those of the Australasian bowerbirds. Males attract their mates with decorated bowers which are a substitute for a colourful courtship plumage. The male satin bowerbird, for example, builds a corridor-like bower decorated with bright blue objects.

are different. The female counteracts the male's aggression with appeasement displays which enable her to remain on his territory. These displays may be similar to the ones used by males when they are submitting to a rival, or may operate by arousing a conflicting tendency in the male. Many female birds, for example, beg for food like a juvenile, thereby reducing the male's aggression by making him feel a desire to feed her. Once a pair-bond has been established, the female's begging results in genuine feeding (called courtship feeding) which not only helps to maintain the bond, but also provides the female with some of the additional food she needs for the production of eggs.

In baboons the females inhibit the aggression of males by directly arousing their sexual feelings. They develop a bright red genital swelling during oestrus and display or 'present' this colourful releasing signal to the dominant male of their group in order to appease and attract him. An interesting development of this form of appeasement is found in male hamadryas baboons. They have bright red, dummy, oestrus swellings and appease higher ranking males by presenting in the same way as sexually submissive females. This inhibits the dominant male's aggression by arousing

Sitting on their small nesting territory, male frigate birds display their enormous red throat pouch to attract passing females. The pouch remains inflated only during the breeding season.

When in oestrus, female baboons develop a red genital swelling which they 'present' to the male in their copulatory invitation display. A similar 'presentation' display is used by both male and female baboons to indicate submission to a higher ranking animal. In the hamadryas baboon, seen here, the males also have red swellings, which mimic those of the females, and so enhance their submissive display.

With their flamboyant plumes streaming behind them, male Jackson's widow-birds bounce up and down like rubber balls on their communal display or lekking grounds. The small dowdy females are attracted to the leks, visiting them each day during the egg-laying period, to mate with the dominant males.

his sexual feelings and sometimes results in the submissive male being mounted. A low-ranking male that is being beaten in a squabble often takes advantage of this form of appeasement by fleeing to a high-ranking male and presenting to him, while continuing to threaten the rival, who is left facing the high-ranking male and must either stop threatening or risk being chased away.

In most species the advantage of threat and courtship displays being conspicuous to distant males and females has to be balanced against the disadvantage of their being conspicuous to predators as well. Threat and courtship displays are therefore most elaborate, and the adornment of the males most flamboyant, in polygynous species in which males are dispensible because only a few are needed to mate with all the females. Polygyny can only occur in species in which the female is capable of caring for her offspring on her own. Though it occurs in several other groups of animals,

polygyny is particularly common in birds, notably such species as peacocks, pheasants, grouse, ruffs, some hummingbirds, cock-of-the-rocks, manakins, birds of paradise, bishop-birds and widow-birds. The males of many of these species congregate on special territorial grounds or leks, where the favoured central positions are occupied by the most successful and dominant males. Because of the competition for space in this area, the central territories tend to be small and rivalry between neighbouring males very intense, providing a strong attraction for the females that visit the lek. As a result, the majority of females go straight to the central area and mate with dominant males. Thus in the sage grouse as many as 400 males may occupy an area half a mile long and two hundred yards wide, but the dominant two or three males at the centre of this area are the only ones actually to mate with females. One 'flockmaster' was seen to mate with 21 hens in the course of a morning.

REPRODUCTIVE ISOLATION

Courtship displays and other signals play an important part in ensuring that animals recognize and mate only with members of their own species, particularly in areas where several similar species live together. An animal's displays usually have some unique feature which is recognized instinctively by other members of the same species. Thus male fence lizards nod their heads with a fixed characteristic rhythm which differs from species to species and attracts only appropriate females. Similarly, each species of fiddler crab has a characteristic way of waving its claw.

However, not all animals recognize each other instinctively, because many lack the simple distinctive markings or other features that make suitable releasers. Female surface-feeding ducks in the genus *Anas* provide good examples; their nondescript, cryptic plumages (essential for safety during breeding) have no really distinctive features and the males have to learn the subtle differences that distinguish the females of their own species. They do this by imprinting on their mother during a period, shortly after hatching, when they are exceptionally sensitive to her appearance. They retain this image or 'imprint' of a mother figure throughout their lives and court and mate only with ducks that conform to it exactly. Male ducklings normally imprint on their mother because she alone is present during the sensitive period, but they can be made to imprint on ducks, or even drakes, of other species if these are substituted for their real mother at the critical time. Thus male mallard ducklings can be made to imprint on a pintail foster parent, with the result that when they are adult they ignore female mallards and court only pintails. On the other hand, it makes no difference if female mallard ducklings are reared by another species; they subsequently respond only to courting male mallards, which they recognize instinctively by their simple, distinctive markings. It should, perhaps, be emphasized that ducklings have no opportunity to imprint on the male, as he takes no part in rearing them. Males must, therefore, have distinctive releasing signals, so as to be recognizable by females and other males.

Sometimes an apparently insignificant colour signal is involved in species recognition. In many northern gulls, for example, the colour of the narrow ring around the eye is critical and individuals probably learn the appropriate colour by observing their parents' eye-rings when they are nestlings. N. G. Smith investigated this method of species recognition by painting the eye-rings of a number of gulls with colours characteristic of others. By doing so he was able to break up estab-lished breeding pairs and prevent unpaired birds from obtaining a mate. By appropriately altering the colour of the eye-rings of male and female glaucous and Thayer's gulls he was even able to induce mixed pairs. However, M. P. Harris found that in the case of lesser black-backed and herring gulls only the females discriminate. Females of one species fostered by the other mated only with males of their foster species. Fostered males were prepared to mate with either species, but because of their own eye-ring colour, were accepted only by females of their own species or by fostered females of the other species.

Among African weavers the colour of the iris may be critical for species recognition, for superficially similar species breeding in mixed colonies are usually found to have differently coloured eyes.

The enormously enlarged and colourful claw of male fiddler crabs is used for threat and courtship signalling and fighting. The courtship signals maintain reproductive isolation, for each species waves its claw in a ritualized and characteristic manner which is attractive only to females of the right species.

In western Uganda, for example, black-headed, yellow-collared and masked weavers are very alike and often breed together, but can be readily distinguished by their respective red, dark brown and straw-yellow eyes.

COLOUR SIGNALS IN THE EGGS AND YOUNG OF BIRDS

Most birds do not need to be able to recognize their own eggs or young, but there are a few notable exceptions. It is, for example, an advantage for species parasitized by cuckoos and other brood parasites to distinguish and discriminate against foreign eggs and young. Such discrimination does occur and sometimes becomes so good that the

Left: Though they behave as distinct species in Europe, the herring (above) and lesser black-backed gulls (below) are very closely related and produce fertile hybrid young when they occasionally interbreed. The two species remain distinct only because they differ in a number of apparently minor characteristics, notably the colour of the ring of skin around the eye, which they recognize.

Even though they breed on crowded rock-ledges, guillemots can find their own egg because it has a characteristic colour and pattern which they have learned to recognize.

parasites are forced to find alternative hosts, in spite of the fact that their eggs and young mimic their hosts' remarkably well.

Another example is provided by species which nest on the ground in dense colonies. Such species usually have variable eggs and locate them partly by recognition of their distinctive colour and pattern and partly by recognition of the nest-site. Guillemots have particularly variable eggs and possess a well developed ability to recognize their own, a great advantage on densely packed rock-ledges where eggs are often rolled about by the wind.

The coloured gape of nestling birds provides a superb example of a very simple releasing signal. Hungry nestlings of many species stimulate their parents to feed them by stretching up and displaying their enormous brightly coloured mouths. Some indication of the strength and simplicity of this feeding stimulus is given by the fact that parent birds will instinctively stuff food into almost any brightly-coloured, gape-shaped cavity. They readily feed appropriate cardboard models and have even been known to feed carp which had learned to open their mouths above the water-surface at the edge of their pond.

Some birds rely on the gaping of their nestlings to distinguish whether they are alive or dead; they feed them if they gape, but throw them out of the nest if they do not. This is a good example of the unreasoning and rigid way in which animals respond to releasing signals. O. Koenig had difficulty breeding bearded tits in captivity because the parents always ejected the young from their nest. He eventually discovered that so much food was being provided that the parents were able to satiate their young, which then stopped gaping. To the parents this meant that they were dead, so they removed them from the nest. Fortunately, this instinctive reaction presents no problem in the wild, because the parents have to gather food from a distance and are never able to satiate their young; any nestling that fails to gape is indeed dead.

Left: The enormous red gape of a young cuckoo (here in a reed warbler's nest) is such a powerful begging stimulus to small Passerine species that birds other than its foster-parents are sometimes stimulated to feed it.

Like the chicks of several other gull species, Dominican gull chicks stimulate their parents to feed them by pecking instinctively at the red spot on their bill.

Parent birds respond most strongly to the largest gape in their brood. This has considerable adaptive value because it ensures that the largest and strongest young are fed first. When foraging conditions are poor small, weak young get little or no food; they soon die, and the brood is quickly reduced to a manageable size. Birds of prey and owls have a well marked runting system of this sort, helped by the fact that incubation begins when the first egg is laid, so that their young hatch at different times and vary in size according to their age. It is far better for a bird to rear a small number of strong, healthy young than a larger number of runts.

Not all young birds beg for food by gaping. Gull and tern chicks, for example, solicit their parents to regurgitate food by pecking at their bill. This pecking response is instinctive and is often stimulated by a colour releaser, such as the contrasting red spot near the end of a parent herring gull's bill. Young ducks and game birds are not fed at all by their parents, but when they hatch they show a marked tendency to peck at green objects, a tendency which may help them to find suitable plant food in their initial feeding efforts. Later, past experience enables them to distinguish edible items.

Disguise and adornment in man

Tribesmen of New Guinea, like the men of many other human societies, decorate themselves with elaborate hair styles, colourful plumes, paint and other ornamentations. Their splendid regalia is intended to impress other males, particularly those of rival tribes, and no doubt wins the admiration of their womenfolk.

Colour adaptations similar to those of other visual animals play an essential part in the life of man. He is not naturally camouflaged, disguised, warningly coloured or even particularly adorned, but his habit of dressing up (which is adventitious adaptation) gives him great flexibility and enables him to adapt to a wide range of situations.

Since time immemorial, man the hunter has been disguising himself in order to get close enough to his prey to kill it. There is a rock-painting of Paleolithic man disguised as a reindeer in the Trois Frères cave in France and similar disguises have been in use all over the world until very recently: North American plains Indians approached herds of bison, which were unafraid of solitary wolves, clothed in a wolf skin; Australian aborigines dressed themselves in emu skins; and various African tribesmen made use of the skins of a great variety of animals, including hartebeestes, zebras and ground hornbills. G. W. Stow described the way in which Bushmen dressed themselves in an ostrich skin to hunt zebra, behaving like an ostrich and using a long stick to keep the head up. 'Most of them were very expert in imitating the actions of the living bird. When they sighted a herd of quaggas which they wished to attack, they did not move directly towards them, but leisurely made a circuit about them, gradually approaching nearer and nearer. Whilst doing so the mock bird would appear to feed and pick at the various bushes as it went along, and rub its head ever and anon upon its feathers, now standing to gaze, now moving stealthily towards the game, until at length the apparently friendly Ostrich appeared, as it was wont in its wild state, to be feeding among them.

'Singling out his victim, the hunter let fly his fatal shaft, and immediately continued feeding; the wounded animal sprang forward for a short distance, the others made a few startled paces, but seeing nothing to alarm them and only the apparently friendly Ostrich quietly feeding, they also resumed tranquillity, thus enabling the dexterous huntsman to mark a second, if he felt so inclined.'

Early photographers used similar methods to obtain photographs of animals. Richard Kearton, for example, photographed many birds with his camera concealed in the head of a stuffed ox.

Man has also made use of camouflage techniques, particularly in modern warfare when killing is carried out at long range. Battledress, guns, tanks, lorries, aircraft and buildings are painted in

Dancing is a form of social display in which one or other, or both of the sexes show off their physical appearance and artistic skill. In this case Australian aborigines on Bathurst Island are holding a dance or corroboree to the music of a didgeridoo.

disruptive colours and camouflage nets are used to conceal shadows. The principles of camouflage were not, however, always fully understood or properly applied during the First and Second World Wars. Cott described the way in which disruptive colouration was misused.

'Various recent attempts to camouflage tanks, armoured cars, and the roofs of buildings with paint reveal an almost complete failure by those responsible to grasp the essential factor in the disguise of surface continuity and of contour. Such work must be carried out with courage and confidence, for at close range objects properly treated will appear glaringly conspicuous. But they are not painted for deception at close range, but at ranges at which big gun actions and bombing raids are likely to be attempted. And at these distances differences of tint—mere blotches of brown and green and grey like those commonly used for 'camouflaging' army vehicles—blend and thus nullify the effect and render the work practically valueless.'

Man is not distasteful and has no use for warning colouration in his relationships with other animals, but he uses the same colours that they use to warn of danger or draw attention to particular objects in

The splendid uniforms of these eighteenth-century Royal Edinburgh Volunteers are typical of those worn by European armies of the time. Their function was exactly the same as the warpaint and plumes of native warriors in other parts of the world—to impress and intimidate the soldiers of enemy armies.

The traditional make-up of Japanese geishas is even more stylized and exaggerated than that of present-day western women, but has exactly the same function.

Far right: Contrasting with the rest of the face, the pink everted lips of humans are a sexual releasing signal associated with kissing. In many societies, women use lipstick to emphasize the colour and glistening texture of their lips and so make themselves more attractive. Similarly, eye make-up is designed to emphasize the eye expression that conveys sexual interest.

his everyday life. Red commonly signifies danger, 'day-glow' orange is widely used as a distress signal and combinations of red and white and black and white are used in warning notices, road signs and road crossings.

Man's social and cultural use of colour is a vast subject and in the space available it is impossible to do more than draw attention to some of the more obvious parallels between man and other animals.

The way in which warriors and soldiers decorate themselves with warpaint, plumes, brightly coloured uniforms and other ornamentations, beat drums, chant, perform war dances, parade and brandish weapons has obvious parallels with the threat displays of other male animals. Military parades and manoeuvres are a modern manifestation of the same sort of display.

The lek displays of birds and other animals find their parallel in the very physical, competitive dances performed by men in many societies. Scottish highland dancing, Morris dancing, Cossack dancing and most African tribal dancing are good examples. There are even cases in which human dances apparently mimic those of lek birds. The traditional dances of the Cree Indians of North America, for example, contain some of the postures

and steps used in the displays performed by male sharp-tail grouse and prairie chickens on their dancing grounds: the birds pivot with rapidly stamping feet, heads down, tails up, and shuffle around together. Some of the Nilotohamitic tribes of Africa, notably the Masai and Samburu, perform competitive jumping dances and it may be more than a coincidence that they live in the same area as Jackson's widow-bird, a lek species which has a remarkable jumping display.

However, most human societies differ from typical lek animals, for in man both the sexes adorn themselves and compete to attract the opposite sex. Adornment of both sexes does occur in some animals, but humans differ from these as well, for they have clear-cut secondary sexual characters and the sexes adorn themselves in different ways; fine clothing is more important to men in some societies and to women in others; the same is true of jewelry and make-up. Fashions of adornment change within limits set by tradition, but tradition itself evolves continuously and sometimes very rapidly if it is subjected to the influence of a powerful outside culture; today Western European and American cultures are having an overwhelming influence on the rest of the world.

Bibliography

BATES, H W: 'Contributions to an insect fauna of the Amazon Valley. Lepidoptera: Heliconidae', *Trans. Linn. Soc., Zool.*, XXIII, pp. 495–566, 1862.

BATESON, W: 'Notes on the senses and habits of some Crustacea', *Journ. Mar. Biol. Assoc.*, I, pp. 211–214, 1889–90.

BELT, T: *The Naturalist in Nicaragua*, London, 1874.

BLEST, A D: 'The function of eyespot patterns in the Lepidoptera', *Behaviour*, 11, 209–256, 1957.

BROWER, L P: 'Ecological chemistry', *Sci. Am.*, 220(2), pp. 22–29, 1969.

BROWER, L P: 'Prey colouration and predator behaviour', in *Topics in the Study of Life*, The BIO Source Book, Section 6, Animal Behaviour, New York, 1971.

BROWER, L P AND BROWER, J V Z: 'Investigations into mimicry', *Nat. Hist.*, 71(4), pp. 8–19, 1962.

BUTLER, A G: 'The larvae of *Abraxas grossulariata* distasteful to frogs', *Ent. Month. Mag.*, V, pp. 131–132, 1868.

BUTLER, A G: 'Remarks upon certain caterpillars, etc., which are unpalatable to their enemies', *Trans. Ent. Soc. London*, pp. 27–29, 1869.

CLARKE, C A AND SHEPPARD, P M: 'Interactions between major genes and polygenes in the determination of the mimetic patterns of *Papilio dardanus*', *Evolution*, 17, pp. 404–413, 1963.

COTT, H B: *Adaptive Colouration in Animals*, Methuen, London, 1940.

DARWIN, C: *Journal of Researches . . . round the World*, London, 1890.

EIBL-EIBESFELDT, I: *Ethology. The Biology of Behaviour*, Holt Rinehart, New York, 1970.

FOX, H M AND VEVERS, G: *The Nature of Animal Colours*, Sidgwick & Jackson, London, 1960.

HARRIS, M P: 'Abnormal migration and hybridisation of *Larus argentatus* and *L. fuscus* after interspecies fostering experiments', *Ibis*, 112, pp. 488–498, 1970.

HUDSON, W H: *The Birds of La Plata*, London, 1920.

KETTLEWELL, H B D: 'Insect adaptations', *Animals* 5, pp. 520–523, 1965.

LACK, D: *The Life of the Robin*, Weatherby, London, 1943.

LAMBORN, W A: 'Further striking evidence of the distastefulness which accompanies the aposematic display of the African Acridian *Zonoceros elegans* Thun.', *Proc. Roy. Ent. Soc. London*, X, p. 4, 1935.

LIMBAUGH, C: 'Cleaning symbiosis', *Sci. Am.*, 205(8), pp. 42–49, 1961.

MARLER, P R: 'Studies on fighting in chaffinches (2). The effect on dominance relations of disguising females as males', *Brit. J. Animal Behaviour*, 3, pp. 137–146, 1955.

MORTENSEN, T: 'Observations on protective adaptations and habits, mainly in marine animals', *Saertryk af Vidensk. Medd. fra Dansk naturhist. Foren.*, 69, pp. 57–96, 1917.

MÜLLER, F: '*Ituna* and *Thyridea*; a remarkable case of mimicry in butterflies', *Proc. Ent. Soc. London*, pp. xx–xxix, 1879.

NELSON, E W: 'The rabbits of North America', *N. Amer. Fauna*, no. 29, *U.S. biol. Surv. Wash.*, 1909.

POULTON, E B: *The Colours of Animals*, London, 1890.

ROTHSCHILD, M: 'Mimicry. The deceptive way of life', *Nat. Hist.*, 76(1), pp. 44–51, 1967.

SMITH, N G: 'Evolution of some Arctic gulls (*Larus*): an experimental study of isolating mechanisms', *Amer. Orn. Union momogr.*, no. 4, pp. 1–99, 1966.

STOW, G W: *The Native Races of South Africa*, Frank Cass 1969, London, 1905.

SWYNNERTON, C F M: 'On the colouration of the mouths and eggs of birds.' *Ibis*, Ser. 10, IV, 4, pp. 529–606, 1916.

SWYNNERTON, C F M: 'An investigation into the defences of butterflies of the genus *Charaxes*', *3rd Int. Ent. Congress, Zurich* (1925), 2, pp. 478–506, 1926.

TINBERGEN, N: *The Study of Instinct*, O.U.P., London, 1951.

TINBERGEN, N: *The Herring Gull's World*, Collins, London, 1953.

TURNER, J G F: 'Passion flower butterflies', *Animals*, 15, pp. 15–21, 1973.

WALLACE, A R: *Darwinism*, London, 1889.

WICKLER, W: *Mimicry*, World University Library, Weidenfeld & Nicolson, London, 1968.

Glossary

aroid Of the arum family of plants.

background picturing The resemblance of an animal in colour and pattern to colours and patterns in its immediate surroundings.

barbels Tactile filaments around the mouth of some fish.

barbules Small filamentous processes (projections) on either side of the larger processes (barbs) which make up the vane of a feather. The barbules of adjacent barbs interlock by means of tiny hooks known as barbicels.

behavioural flattening Flattening, brought about by an animal's behaviour, that results in its shape merging with the ground or other surface.

carnivore In a general sense, any animal that feeds on other animals; in a restricted sense, an animal that belongs to the order Carnivora.

carotenoids A group of plant pigments that are yellow, orange or red.

cervical Appertaining to the neck.

chela The pincer of a crab, lobster or other crustacean.

chitin The principal constituent of the exoskeleton (the hard external covering) of insects and other Arthropods.

chromatophore A cell containing pigment. By changing the distribution of pigment in their chromatophores some animals are able to change colour.

countershading The way in which many animals are shaded darker above and paler below to counteract shadows caused by natural overhead light.

cryptic colouration Any colouration that contributes to the concealment of an animal.

differential blending A form of disruptive colouration in which part of the colour pattern of an animal merges with its surroundings; the contrasting part gives no indication of the shape of the animal.

diffraction The splitting of a ray of light into dark and light bands or spectral colours when one ray interferes with another as it is deflected at the edge of an opaque object or passes through a narrow slit.

digitalis A poison found in the leaves of foxgloves. It is the source of drugs used in the treatment of congestive heart failure.

disruptive colouration Boldly patterned colouration that breaks up the shape of an animal, thereby making it less conspicuous.

herbivore Any animal that feeds on plants.

imprinting A very rapid, irreversible form of learning whereby newly hatched or newborn animals learn to recognize their parents.

inflorescence A group of flowers more or less sharply defined from the purely vegetative parts of the plant.

instar A stage in the larval development of an insect between two successive moults.

interference of light A process in which colours are produced by physical structures; the process depends on the interaction of out of phase rays of light which results in certain wavelengths being reinforced or neutralized.

intergrade The way in which categories sometimes merge imperceptibly without clearly defined limits.

keratin The principal constituent of various epidermal structures in vertebrates, notably hair, horn, nails, claws, feathers and scales.

leguminous Appertaining to plants in the Leguminosae, a family that includes peas and beans.

lek A communal ground or arena where the males of some bird species assemble to display. Females visit the lek and mate promiscuously with the males.

melanins A group of animal pigments that are generally black or brown, but sometimes red or yellow.

melanism Excessive blackness of scales, feathers, hair or skin due to an overabundance of the pigment melanin.

melanophore A chromatophore containing **melanin**.

natural selection The preferential survival of the best adapted varieties or mutants, resulting in evolutionary change.

nematocysts Specialized cells of jellyfish, hydroids and other Coelenterates that eject a stinging or sticky thread when triggered by the touch of another animal.

nymph An insect larva that resembles an adult but for the fact that the wings and reproductive organs are undeveloped.

petiole A leaf stalk.

polygyny The form of polygamy in which males mate with more than one female.

polymorphism The occurrence of several different forms in the same species.

Polyzoa A group of mostly marine invertebrates that live colonially and propagate by budding.

preen gland A gland, situated at the base of the tail of birds, that secretes oils used in feather maintenance, particularly waterproofing.

pterins A group of pigments, found mainly in insects, that are bright yellow or orange.

releaser A stimulus which sets off an instinctive behaviour pattern.

saturated In this sense, appertaining to colours that are undiluted by white.

searching image A mental image of prey that an animal carries while hunting. As a result of its searching image the animal hunts more efficiently for the particular prey concerned, but tends to overlook alternative prey.

sessile Being stationary and more or less fixed to the substrate, as in sponges, corals, sea anemones and barnacles.

spicules The often needle-like bodies, made of silica or calcite, that form the endoskeleton (the internal framework) of sponges and some other invertebrates.

spiracles The external openings of the tracheae (breathing tubes) of insects.

stroma The tissue which forms the ground substance of an organ.

structural flattening Flattening, brought about by an animal's structure, that results in its shape merging with the ground or other surface.

thoracic Appertaining to the thorax.

urticating Having an effect like that of a nettle sting; producing inflammation and itching.

Index